PATHWAYS TO OUR SUSTAINABLE FUTURE

The Duquesne Incline was one of four inclines serving the summit of Coal Hill, which later came to be known as Mount Washington. From this view one can see why Pittsburgh was designated a Biophilic City in September 2016.
CITYSCAPE PHOTOGRAPHY

PATHWAYS TO OUR SUSTAINABLE FUTURE

A PERSPECTIVE FROM PITTSBURGH

PATRICIA M. DEMARCO

UNIVERSITY OF PITTSBURGH PRESS

Published by the University of Pittsburgh Press, Pittsburgh, Pa., 15260

Manufactured in the United States of America
Printed on acid-free paper
10 9 8 7 6 5 4 3 2 1

Cataloging-in-Publication data is available from the Library of Congress

ISBN 13: 978-0-8229-6501-5
ISBN 10: 0-8229-6501-1

Cover photograph: H. Mark Weidman Photography / Alamy Stock Photo
Cover design: Joel W. Coggins

This book is dedicated to the children
of the twenty-first century.

CONTENTS

ACKNOWLEDGMENTS

THIS BOOK PRESENTS the work of many people, companies, and institutions moving toward a sustainable future through the inertia of the industrial age. Pioneers in sustainability have been addressed in many other works, but the struggle for sustainability in a city whose traditional identity is so bound in the fossil fuel age has special importance. I am thankful for their stories and their part in the process of shaping change in this time and place—Pittsburgh.

I have special thanks for Jeanne Pearlman for challenging me to write the book in a defined time, and providing the resources to make that possible. The support from the Pittsburgh Foundation W. Clyde and Ida Mae Thurman Fund allowed me to focus on writing as my primary pursuit and was the catalyst for this work. I am grateful for the constant support and encouragement, expertise, and critical discussions with Terry Collins, Teresa Heinz Professor of Green Chemistry, who gave me an institutional home at Carnegie Mellon University. Linda Lear has served as a mentor and friend throughout this process.

My close colleagues for critical discussions and readings have shaped the tone and helped with reviews for accuracy and context. In many ways, this has been a shared journey with my closest colleagues: Stan Kabala, Mark Madison, John Stolz, Mark Collins, Matt Mehalik, Maren Cooke, Robert Mulvihill, Margaret Zak, Paula Purnell, Volker Hartkopf, Vivian Loftness, Marc and Christine Mondor, Cliff Davidson, Lavern Baker Hotep, John Rohe, Deborah Rohe, Dave Hassenzahl, Mark Dixon, Kirsi Jansa, Anne Rosenthal, Alice Julier, Robert Sroufe, Joel Tarr, David Carlisle, Robin Alexander, Elisa Beck, Patricia Lemer, Tom Peters, Charles McCollester, Mike Stout, and Rosemary Trump. Marie Fechik-Kirk has been a critical help in this effort, serving as my research assistant and helping to photograph and document our interviews. To Portia "PK" Weston for her astute developmental editing, I owe a debt I can never repay.

I am grateful to all of the people who shared their expertise, stories, and resources for this book: Brenda Smith, Judy Focareta, Louis Hinds, Ron Gdovic, Fred Kraybill, Richard Piacentini, Hal Saville, Michael Carnahan, Jack Scalo, Mayor Bill Peduto, Bill Generette, Congressman Mike Doyle, Grant Ervin, Sara and Joseph Bozzelli, James (Jamie) Moore, Don Kretschmann, Shelly Danko-Day, Michelle Naccarati-Chapkis, Chelsea Holmes, Colin Horowitz, Lalit Chordia, Eric Beckman, Bill O'Rourke, Ned Eldridge, Andrew Butcher, Rhonda Sears, LaKeisha Wolf, Jeaneen Zappa, and Fred Brown.

The logistics and mechanics of converting thirty years of life experience and six years of teaching into a coherent and interesting text were daunting hurdles. I am eternally grateful to Bernie Goldstein for introducing me to my literary agent, Joan Parker. To my editor Sandra Crooms, Amy Sherman, and the production team at the University of Pittsburgh Press, I am grateful for the guidance and professional polish added to my work.

This book was many years in gestation and would never have come to print without the encouragement of my family: Michael and Maryann DeMarco, Linda DeMarco Miller and Randy Miller, and Tom Jensen for his patience during the writing. This book is for the children whose future we are shaping with our actions. I am grateful beyond measure for the children in my life: my son Steven Michael Smith and my daughter Jennifer Claudine Griffith, whose talent, enthusiasm, and joy in living have held me through the toughest times. I am grateful for the children of the next generation: my grandchildren, Quinn Harper Griffith and Cecelia Leta Griffith; the children of my nieces and nephew, Julia DeMarco, Mikhail DeMarco, Isaac Avondolio, Stella Avondolio, and Miles Avondolio; and for Tom's grandchildren Kristin Cunzolom, Teagan

D. Jensen, and Peyton Kazmarski. I am grateful for the inspiration and hopefulness I receive from my students at Pitt, Duquesne, and Chatham. The future rests on their shoulders.

I am grateful to acknowledge the generous contributions of the many people who have contributed their thoughts, critique, and reviews of this manuscript. In spite of all efforts, any errors that remain are my responsibility alone. I welcome the thoughts and comments of readers and students, and look forward to an ongoing dialogue through my blog at www.patriciademarco.com.

FOREWORD

IN ***PATHWAYS TO OUR SUSTAINABLE FUTURE,*** political scientist and environmental policy expert Patricia DeMarco has given us a deeply thoughtful analysis delineating why we have ignored unpleasant environmental markers of pending disaster, and how we just might save ourselves *from* ourselves in the twenty-first century.

Hoping to engage in a broader public dialogue, DeMarco brings a lifetime of experience to her task. Her work with public and private organizations and her deep knowledge of Rachel Carson's environmental ethic contribute to the breadth of possibilities she outlines here.

DeMarco has also taken Rachel Carson's holistic vision to a new level of application: insisting that we cannot afford to wait any longer to face our looming environmental disaster. *Pathways to Our Sustainable Future* argues with compelling examples that maintaining the economic and political status quo is no longer an option.

Carson—most famous for her exposé in *Silent Spring* (1962) of the systemic harm done by the widespread misuse of synthetic pesticides—was first and foremost an ecologist. DeMarco rescues Carson from the label of angry crusader by exploring her broader ecological vision most eloquently expressed in her trilogy of books on the sea—particularly in her best-seller *The Sea Around Us* (1952). Carson's understanding of the dynamic rhythms and ecology of the oceans enabled her to articulate a larger environmental ethic, which, if practiced, would embrace and sustain the whole of the living world.

DeMarco follows Carson's vision, offering her own analysis of the many "pathways" open to us to create a sustainable future—an ethic built on cooperation rather than exploitation. She offers fascinating case studies illustrating ways in which individuals and communities can apply the latest "green science" to help solve seemingly insurmountable problems; ways in which self-interest can be transformed into incentives to bring about a larger change in consumer awareness and demand.

Using the environmental history of the city of Pittsburgh, Pennsylvania, DeMarco offers specific examples of the ways in which people and communities can adapt to a planet running out of resources. Here is a seminal, beautifully written analysis of what is wrong. But even more, it is a realistic exploration of how the living world can still be preserved for our children and our grandchildren.

Linda Lear, author of *Rachel Carson: Witness for Nature* (2009)
Bethesda, Maryland

PATHWAYS TO OUR SUSTAINABLE FUTURE

INTRODUCTION

The contamination of the environment with harmful substances is one of the major problems of modern life. The world of air and water and soil supports not only the hundreds of thousands of species of animals and plants, it supports man himself. In the past, we have often chosen to ignore this fact. Now we are receiving sharp reminders that our heedless and destructive acts enter into the vast cycles of the earth and in time return to bring hazard to ourselves.

RACHEL CARSON

THREE ICONIC BRIDGES in Pittsburgh mark great heroes who represent the city's ethos in art, sports, and innovation—artist Andy Warhol, baseball player Roberto Clemente, and the environmental philosopher Rachel Carson. Carson has been recognized as one of the most important thought leaders of the twentieth century.[1] Her eloquent exposition of the inner workings of the oceans that provide us with oxygen, nutrients, and water endeared her to millions. In a day when the written word was the most prominent form of thought-provoking discourse, Carson's influence through her book *Silent Spring* reached the highest levels of government and evoked the concerns of millions of people who read and believed her words. In the fifty intervening years since *Silent Spring* was published, her call for precaution in protecting the living earth and the need to preserve the interconnected web of life, of which humans are but one part, has continued to strike against the mainstream approach of continued reliance on extractive industries. Carson's insistence on placing nature, rather than human convenience and profit, at the center of decisions about resource management assures her continued place as a revolutionary. Her environmental ethic, forged from growing up in Pittsburgh as her town shifted from agrarian to industrial, offers guidance in this new transition to a future based on preserving the natural world.

In the twenty-first century, the cumulative consequences of human behavior now threaten the viability of all life on earth. Earth bears the scars from 150 years of industrialization, which brought about economic progress without much regard for polluting the air, water, and land.[2] Today the long-term consequences have arrived in the form of climate change and global pollution. Climate change compromises the life support system of all oxygen-breathing, freshwater-dependent organisms, including humans, while global contamination from synthetic chemicals and their byproducts, especially those with endocrine-disrupting properties, threatens the health of creatures throughout the biosphere.[3] If the children of the twenty-first century are to have a viable future, human destruction of the interconnected web of life must stop. The window of opportunity to change direction may be very narrow, the remaining time to take a corrective course very brief.

To transition quickly from an economy that depends on fossil material—oil, gas, and coal; synthetic chemicals—to a system that operates on a renewable and sustainable base requires a commitment to preserving the living, natural world. The challenge of this transition is enormous because all aspects of modern civilization depend on fossil fuels. Transportation, power, heat, light, and machinery all operate on petroleum, coal, and natural gas. Industrialized food production depends on petrochemically derived fertilizers, herbicides, pesticides, and fossil-fuel-powered mechanization for cultivation, transportation, processing, and distribution. Construction materials, textiles, packaging, and consumer products come from plastics and synthetic materials derived from fossil carbon sources. The entire economy runs on the infrastructure of extracting, transforming, and consuming fossil carbon, which compromises the viability of global living systems upon which future generations will depend. However, making a shift away from the fossil-fuel-based economy is critical for many reasons, including and especially climate change, wasteful and inefficient energy production processes, national security, and pollution and its effects on human and environmental health.

To set the discussion of climate change in context, it is important to recognize that the shape of the earth's orbit around the sun is the greatest long-term determinant of the earth's climate. As first described by the Serbian astrophysicist Milutin Milanković (1879–1958), the orbit of the Earth slowly shifts from an eccentric elliptical path to a nearly circular path over a period of 90,000-to-100,000-year cycles.[4] When the orbit is most elliptical, the winter period when the Earth is farthest from the sun lasts much longer than the summer period, allowing ice to accumulate progressively, forming the Ice Ages. When the orbit is more nearly circular, the climate is more temperate. Milanković also described the tilt of the Earth on its axis, creating the seasonal variations in the amount of solar exposure, and the "wobble" of the Earth's rotation on its axis. The current orbital configuration is closer to circular, Holocene epoch, and expected to last for approximately another ten thousand years before the elongated orbit ushers in the next major period of glaciation.[5]

Within the Holocene period—roughly aligned with the current orbital configuration and beginning approximately 11,700 years ago—the climate has been controlled by three major forces: the periodic oscillation of the sun's intensity, cycling over an eleven-year pattern of waxing and waning; the alternation of warming and cooling from El Niño and La Niña in the ocean heat sink in deep waters; and the periodic eruptions of volcanoes. These forces have shaped the variation in weather and the overall pattern of climate cycles for millennia, as documented by empirical data from ice cores and fossil sediments.[6] Over ten thousand years, the average period between glacial stages where ice covers the earth, the combined effect of all of these forces cycled the climate within a range of 12 degrees Celsius.[7] However, within the last hundred years, and most acutely in the last fifty years, a fourth influence on the climate has taken place: human activity.[8] Burning fossil fuels for energy is the most significant cause of climate change in modern times.

The steady, temperate climate with an oxygen-rich atmosphere that has characterized the Holocene epoch is shifting. Examined objectively, the evidence is alarming. The global carbon dioxide concentration has escalated sharply in the last fifty years from its preindustrial level of 278 parts per million (ppm) to over 400 ppm in September 2014, with an expected rate of rise of 1.8 ppm per year.[9] The atmospheric carbon dioxide concentration now is higher than has occurred at any time in the last fifteen million years, according to paleoclimatic and geologic evidence.[10] The science is unequivocal that humans are the cause of global warming, and major changes are already being observed: Global mean warming is 0.8°C above preindustrial levels; oceans have warmed by 0.09°C since the 1950s and are acidifying; and sea levels have risen by about 20 centimeters since preindustrial times and are now rising at 3.2 centimeters per decade. An exceptional number of extreme heat waves have occurred in the last decade, and major food-crop-growing areas are increasingly affected by drought.[11] In addition, the cycles from hot to cold are more severe, with lengthy drought cycles and winter arctic vortex conditions dipping far into the continental United States and lingering for days.[12]

Compared to the early industrial period, scientists now have a better understanding of the connections between fossil fuel combustion adding billions of pounds of carbon dioxide to the atmosphere and the changes in the composition of the air, in climate, and in the composition and chemistry of the waters of the ocean.[13] According to a 2011 report by the National Research Council,

> the scientific consensus shows conservatively that for every degree of warming, we will see the following impacts: 5–15 percent reductions in crop yields; 3–10 percent increases in rainfall in some regions contributing to flooding; 5–10 percent decreases in stream-flow in some river basins, including the Arkansas and the Rio Grande, contributing to scarcity of potable water; 200–400 percent increases in the area burned by wildfire in the US; 15 percent decreases in annual average Arctic sea ice, with 25 percent decreases in the yearly minimum extent in September.

Even if all carbon dioxide emissions stopped, the climate would continue to warm for several more centuries. Over thousands of years, this could unleash amplifying feedbacks leading to the disappearance of the polar ice sheets and other dramatic changes. In the meantime, the risk of catastrophic wild cards "such as the potential large-scale release of methane from deep-sea sediments" or permafrost, is impossible to quantify.[14]

America has not yet seized climate change as a galvanizing issue for leadership. Political debate often touts the lack of certainty about the causes, permanent effects, or human contributions to climate change. The US military, however, recognizing its responsibilities for preparedness regardless of political consensus, has responded with precautionary planning. In his introduction to the Department of Defense's climate change analysis, the Climate Change Adaptation Roadmap, published in 2014, then-secretary of defense Chuck Hagel states, "Politics or ideology must not get in the way of sound planning. Our armed forces must prepare for a future with a wide spectrum of possible threats, weighing risks and probabilities to ensure that we will continue to keep our country secure."[15] Secretary Hagel notes, "The challenge of global climate change, while not new to history, is new to the modern world. Climate change does not directly cause conflict, but it can significantly add to the challenges of global instability, hunger, poverty and water shortages, pandemic disease, disputes over refugees and resources, more severe natural disasters—all place additional burdens on economies, societies and institutions around the world."[16]

Air pollution is another inevitable consequence of coal combustion in factories or power plants, oil refining and gasoline combustion in cars, or natural gas extraction. As the cumulative effects of fossil fuel combustion continue to increase, the health of people everywhere is showing the strain. In announcing new proposed regulations for reducing emissions from existing fossil-fueled electric generation units, EPA administrator Gina McCarthy said, "Rising temperatures bring more smog, more asthma, and longer allergy seasons. If your kid doesn't use an inhaler, consider yourself a lucky parent, because one in 10 children in the U.S. suffers from asthma. Carbon pollution from power plants comes packaged with other dangerous pollutants like particulate matter, nitrogen oxides, and sulfur dioxide, putting our families at even more risk."[17] In addition to carbon dioxide and the fine particulate matter, sulfur dioxide, and nitrogen oxides mentioned by McCarthy, the air we breathe is now also infused with volatile organic compounds, heavy metals, and radioactive materials released from the mineral deposits. A 2010 study by the Clean Air Task Force estimated that air pollution from coal-fired power plants accounts for more than 13,000 premature deaths, 20,000 heart attacks, and 1.6 million lost workdays in the United States each year. The total monetary cost of these health impacts is over $100 billion annually.[18] The oldest 10 percent of coal-fired power plants, those operating for fifty years or more, contribute the most emissions, and cause the majority of the health and climate effects.[19] Another aspect of environmental pollution having a significant impact on human health is the production and use of synthetic chemicals and pharmaceuticals. For example, many synthetic chemicals affect the endocrine system, the internal communication system of all animals, including humans. Natural hormones within the body bind to a specific receptor to produce a specific response. The timing and duration of the hormone's presence can drastically change the effects. Some synthetic chemicals in common use today can act like hormones, in that they can bind to the hormonal receptor and block the normal hormone from acting, or they might mimic the natural hormone but cause an abnormal response.

As people accumulate exposure to these endocrine-disrupting contaminants, the normal endocrine system comes under stress. The chemical messengers that coordinate the functions of the brain, the immune system, and the endocrine system operate in the body at extremely low levels. Most natural hormones are active in blood concentrations of parts per trillion. That would be equivalent to a drop of dye in a train of railroad tankers ten miles long. Thus, even if such endocrine-disrupting chemicals appear in very low concentrations in our environment—parts per billion, for example—that is still much higher than the active level of natural hormones in the body. Numerous studies show that astoundingly small quantities of these hormonally active compounds can wreak all manner of biological havoc, particularly on those exposed in the womb.[20]

The rise of environmental awareness in the early 1970s sought to legislate air and water emissions, and limit the use of some kinds of chemicals, but this regulatory approach has allowed billions of pounds of toxic materials into the environment legally, without addressing the underlying problems. For example, approximately 3.89 billion pounds of toxic chemicals were released in the air or water or were disposed in landfills in 2014.[21] Even if emissions are controlled to a very low concentration, if the compound or its byproducts are taken into the food chain they can become concentrated. Consequently, very low levels in water or soil can become more dangerous to tertiary consumers, such as people or eagles, at the top of long food chains.

PATHWAYS TO A SUSTAINABLE FUTURE—WHY PITTSBURGH?

The challenges manifest today present a daunting complex of causes and effects across all aspects of society, but an examination of the progress of a single important city may provide insight for a way forward.

Pittsburgh, Pennsylvania, sits at the confluence of the Allegheny and the Monongahela Rivers forming the Ohio, which flows into the Mississippi River and, ultimately, the Gulf of Mexico 1,981 miles away. An attractive center for civilization and trade among indigenous peoples who occupied the land for centuries before European colonists arrived, and contested by the French and the British for its strategic location, Pittsburgh played an important role in the formation of America as an independent nation. The two rivers funneled both travelers and materials into the growing city as it became a center for manufacturing, trade, and westward migration. In the late 1800s Pittsburgh led the Industrial Revolution in mobilizing and organizing both labor and industrial capacity.[22] During its peak, post–World War II Pittsburgh personified the hardworking toughness that won the war and built the industrial might of the country.

Pittsburgh's glass, metal, and chemical industries built the nation with a flourishing spirit of innovation, invention, and grit. And just as Pittsburgh was a leader in the rise of the industrial age, this city has taken a lead in the shift to the modern era of the electronic, high-tech-based industry of the future. The transition catalyzed by the collapse of the local steel industry in the late 1980s grew from the tremendous intellectual capital of Pittsburgh. The city's concentration of prominent universities attracts talent and resources, and with a strong collaborative spirit and leadership, the universities have made major contributions to innovation and problem solving both in industry and government. The people of Pittsburgh have pride, resilience, a strong work ethic, and a sense of place.

This is also a time of sharp division with the city and the surrounding areas with people's goals and expectations moving in different directions. Pittsburgh stands at an intersection between the historic extractive and fossil-fuel-based industries and the potential for a renewable and sustainable future. The advance of the Marcellus and Utica shale gas discoveries draws a tremendous amount of resources and attention to the region. County, state, and federal laws and policies strongly support the shale gas industry. But between and under this surge of the newest fossil fuel development, a different path emerges within the interstices. Active struggle from differing perspectives and motives fuels the debate about how best to move forward, yet out of this struggle a new vision for Pittsburgh has emerged, one built on sustainability and resilience principles, respect for the diversity of cultures, and precaution in restoring and preserving the health of its natural environment.

Pathways to a Sustainable Future shows how Pittsburgh is addressing sustainability issues. The solutions are not necessarily found in technology alone; rather, the pathways forward are based on the ethical and moral basis for making choices about the future. The case studies presented here show the decision points scattered throughout the fabric of this community where individuals, companies, and institutions have made decisions to move in a new direction, not necessarily in the mainstream.

> *Those who contemplate the beauty of the Earth will find reserves of strength that will endure as long as life lasts. There is something infinitely healing in the repeated refrains of nature—the assurance that dawn comes after night, and spring after winter.*[23]
>
> **RACHEL CARSON**

Pathways to a Sustainable Future travels in the direction Rachel Carson indicated—a path to where people live in harmony with nature's laws, as participants in thriving ecosystems. The first part addresses the essential connection between people and the earth, its living systems, and how people can reconnect with the natural world. The second addresses some sustainable directions to providing energy, food, and materials. It explores the policies embedded in the current system and some of the new directions to achieve change. The third part examines the social and cultural impediments to change and the means to address them. This is a book of empowerment, inspired by Rachel Carson whose voice rose in challenge to a system that presented danger to living systems and moved thousands to respond. The situation facing humanity today calls for a unified response in defense of the earth, for the sake of all of its living creatures, including the children of the twenty-first century. The stories of Pittsburghers blazing new pathways share their struggles and their hopes for a sustainable future. It is time to act.

View of Pittsburgh's skyline from the west end.

TREE PITTSBURGH

PART I

CONNECTING TO THE LIVING EARTH

The stream of time moves forward and mankind moves with it. Your generation must come to terms with the environment. Your generation must face realities instead of taking refuge in ignorance and evasion of truth. Yours is a grave and a sobering responsibility, but it is also a shining opportunity. You go out into a world where mankind is challenged, as it has never been before, to prove its maturity and its mastery—not of nature but of itself. Therein lies our hope and our destiny.

RACHEL CARSON, COMMENCEMENT ADDRESS TO SCRIPPS COLLEGE, CLAREMONT, CA, JUNE 12, 1962

A clutch of hungry baby robins in a nest at the author's home in Pittsburgh.
PATRICIA DEMARCO

1

LISTENING TO THE VOICE OF THE EARTH

In spite of the truly marvelous inventiveness of the human brain, we are beginning to wonder whether our power to change the face of nature should not have been tempered with wisdom for our own good, and with a greater sense of responsibility for the welfare of generations to come.

RACHEL CARSON

WESTERN CIVILIZATION HAS considered the conquest of nature as a major accomplishment. In the twentieth century, technology conquered many frontiers set by physical limits of land, space, and the biological cycles of life. The scourge of disease and illness has been abated through modern advances in medicine to extend life expectancy and reduce infant mortality. The power unleashed from splitting atoms has produced weapons of warfare and also nuclear-powered electricity. The dream of humans setting foot on the moon has become reality. The genetic code has yielded its secrets to scientists who can now clone plants and animals and create genetically modified organisms of human design. These and many other technological innovations have made the way of human development easier, more convenient, less dangerous, and less prone to illness. But the triumph of humans over nature will be a hollow victory if the result is an uninhabitable world. Many of these achievements have come at the sacrifice of natural environments.[1] Thus, within 150 years of the Industrial Revolution, resource use without replacement, restoration, or protection has precipitated global crises in water availability, air quality, and land use. Human activities are changing the geochemical balance of the planet.[2] Threats from climate change and global pollution challenge everyone in the twenty-first century to reexamine how humans relate to the natural environment. To move civilization forward, people must recognize our place in the living world.

Many times over its history, the land has changed at the hands of human enterprise. In precolonial days, indigenous peoples expanded their civilizations through resource management systems that supported vast economies with intracontinental trade involving millions of people.[3] For thousands of years indigenous peoples all over the world thrived through cultures based on a reverence for nature. On the American continent, the earliest European colonists came to a land that had been occupied for centuries by Native American tribes, with members numbering about four to ten million people at the beginning of the seventeenth century.[4] Native peoples lived using modest clearing techniques to create and maintain grazing lands. They foraged and hunted, often in nomadic patterns that cycled with movement of prey, and they held property in common to use but not to possess. European concepts of land ownership and "improvement" starkly contrasted with these permaculture and modestly intrusive practices. With advanced weaponry and the decimation of Native populations by European diseases, the traditions that had sustained abundance for centuries were overrun by practices that revolved around land ownership and resource exploitation. Indeed, the colonial intent was to extract resources from the new colony to supply the mother country.[5]

Over the last two hundred years coal mining has left a legacy of polluted rivers and streams and permanently scarred land. In Pennsylvania, over three thousand miles of waterways are polluted with acid mine drainage.[6] Even after mines are closed, the water running through them creates a continuing stream of rust-colored effluent uninhabitable for normal aquatic ecosystems.[7] In West Virginia, one-third of all rivers and streams are reported as sterile of living creatures downstream of mountaintop removal mining.[8] As the deeper coal seams are played out, mining companies reach thin seams by removing the tops of the mountains, blasting the "overburden," and dumping it into the valleys. The flattened hills stand like moonscapes of exposed rock banked with roads where the coal was removed.

Resource-extractive practices like mountaintop removal coal mining obviously devastate the immediately affected areas, but less deliberate human actions can have equally destructive effects. As housing developments, agriculture, road building, and cities expand, nonhuman species and natural land uses are crowded out or destroyed, forming holes in the web of life. Commercial plantations for palm oil, for example, destroy critical habitats such as tropical rain forests.[9] Human activities, primarily those driving habitat loss, the introduction of exotic species, and global warming, place 99 percent of currently threatened species at risk.[10] An estimated thirty thousand species are going extinct every year—a rate of three species per hour, compared to a background rate (absent human influence) of one extinction event per million species per year.[11]

Extinction is not an isolated event. Its effects cascade as ecosystems unravel because of the complex interconnections among other species. The accelerating extinction of species is one of earth's distress calls, a signal that the balance between the living and mineral elements of the earth that evolved over millions of years is shifting. While it is true that conditions on earth have changed dramatically over geologic time, the rapid pace of change evident today has no precedent.[12] The cumulative effects of industrialization have compromised the life-supporting composition of the atmosphere and the geochemical systems it controls—surface temperature, water cycles, and ocean acidity. Living systems rarely adapt to rapid changes in living conditions, especially those such as plants rooted in place with long regenerative life cycles or ocean communities subjected to changes in the chemical composition of seawater. This is an ethical challenge of global scale and lies at the heart of the rationale for acting to correct the human disruptions of natural living systems.

The complex interconnections among living things form Earth's life support system, necessary for all of today's creatures and for future generations—fresh air, clean water, fertile ground, and the biodiversity of species that make up the interconnected web of life. The challenges of climate change and chemical contamination present a call to preserve the living earth. It is a call to temper the prowess of technology with wisdom and precaution to protect Earth's living systems. It is a plea for justice for those who will be most acutely affected, those whose voices are not included in the debate, and those who are disproportionately vulnerable. It is a plea for accountability in the way people have used the natural resources of the earth for short-term benefits. It is a plea for life to exist.

THE ETHICAL CHALLENGES OF CLIMATE CHANGE

> *Global climate change poses a clear and immediate threat to the well-being of the community of life on Earth, which includes non-human life, but also the social and cultural dimensions of human existence. However, while it is clear that climate change deserves focused ethical consideration and intervention, vulnerability seems poorly understood.*
>
> **UNESCO WORLD COMMISSION ON THE ETHICS OF SCIENTIFIC KNOWLEDGE AND TECHNOLOGY**

Facts and technology alone will not yield clear lines of consensus for action; rather, recognizing the consequences of human actions with empathy for the effects on other people and other living things compels action. The need for an ethical rather than a strictly technical approach to finding solutions is predicated on four reasons: the global dispersion of the causes and effects of climate change; the fragmentation of agency; institutional inadequacy; and the persistence, nonlinearity, and time-delayed nature of climate change.[13]

Climate change was first brought forward as a global issue in the first Intergovernmental Panel on Climate Change Report in 1990.[14] Since that time, there have been four IPCC assessment reports, each calling for action with increasing urgency. Yet all of the Conferences of Parties that have convened international gatherings of nations to address the issue of climate change have fallen short of the goal. The twenty-first Conference of Parties in Paris (COP21) in December 2015 achieved commitments from 195 nations, but even if all promised goals are met, the expected impact on containing global warming will fall far short of what is required.[15]

The global nature of the effects of climate change defines an ethical situation where each nation experiences the harms of climate change, regardless of its contribution to the causation. This may be understood as a problem of a global commons: "In resource terms, the atmosphere is an example of a global commons, a status it shares with the oceans and seabed, outer space, and Antarctica. A global commons may be loosely defined as a domain that is beyond the exclusive jurisdiction of any one nation but one that all nations may use for their own purposes (such as extracting resources or discharging pollutants). Without effective controls, the use of a commons may increase to the point that it becomes severely depleted, contaminated, or degraded."[16]

Climate change and global pollution have multiple sources, from disparate locations. Current agreements depend on voluntary actions from nations who accept no external review or enforcement by sanctions for failure to meet commitments. Unfortunately, regardless of the source of greenhouse gas emissions, the climate effects are experienced globally.[17] This has been described as a global "tragedy of the commons," wherein nations collectively recognize that cooperation to restrict overall pollution and greenhouse gas emissions is in the best interest of all, but at the same time, each nation individually wishes to limit restrictions on its own emissions.[18]

Moreover, while the effects of greenhouse gas emissions are a global phenomenon, they are not evenly distributed. Emissions are distributed by dispersion through the atmosphere and by the cycle of water from the atmosphere to the rivers and oceans. The natural laws of physics and chemistry direct the behavior of materials released into the natural environment; thus, the effects of emissions do not fall uniformly on the different countries of the world—and, more pertinently to a discussion of ethics and justice, these harmful effects are not felt exclusively by those who caused them. Questions of fairness arise as less-developed countries see even more severe constraints on their own resources.

An analysis of greenhouse gas emissions by income level and by country determined that the richest 10 percent of the population accounts for 49 percent of the greenhouse gas emissions while the poorest 50 percent of the world's population accounts for only 10 percent of global greenhouse gas emissions.[19] The disparity becomes more acute as the effects of climate change include rising ocean levels, affecting about one billion people in coastal communities and island nations, many of whom have among the lowest carbon emissions.[20]

Already the consequences of industrialization are coming due, and it is clear that those experiencing the worst of the effects are not the ones who caused them. At COP21, which brought together nations seeking consensus on a way forward to control global greenhouse gas emissions, the debates among factions and representatives from different geographic areas brought the ethical issues at the heart of the dispute into sharp focus.[21] Countries that have been exploited for resources and those located at sea level challenged the moral and ethical responsibility of countries that have benefited economically from their plight. With a plea to prevent sea level rise from inundating their lands, a coalition of Pacific Island nations called for stricter limits on the target for climate control—to 1.5 degrees Celsius (2.7 degrees Fahrenheit) above pre-industrial levels, rather than 2 degrees Celsius (3.6 degrees Fahrenheit).[22]

Addressing such disparities raises the question of who has responsibility for taking action. There are precedents for international collaboration in restricting a recognized environmental threat. For example, the 1987 Montreal Protocol on Substances that Deplete the Ozone Layer achieved replacement of ozone-safe substitutes for chlorofluorocarbons (CFCs) as refrigerants and propellants. In this case, developed nations took the lead in the actions for recovery, and funded the technology transfer for global adoption of the changes.[23] This is a significant demonstration of the principle that the cost causer should be the cost payer, widely recognized in regulatory regimens as a basic tenet of fairness.

Nevertheless, transferring this successful process to the problem of climate change has proven more difficult. In the case of CFCs, the same industries that produced the ozone-depleting chemicals developed the substitutes and benefited economically from the replacement with new materials. In the case of climate change, the source corporations for fossil fuels and potential replacement sources from renewable energy systems are not necessarily aligned. In fact, the distributed nature of renewable resources makes the institutional infrastructure of the fossil industry singularly unsuited to the transformation needed.

Distrust among nations and lack of any effective international enforcement powers has complicated response to an ever more urgent situation. In essence, addressing global climate change appears to require global regulation of greenhouse gas emissions, including a reliable enforcement mechanism, but the current global system—or lack of it—makes this difficult, if not impossible.[24] The laws that govern individual states or countries focus on localized conditions, but have no authority over global operations of multinational corporations, for example.

In the absence of a universally accepted authority to compel cooperation, the situation continues unresolved. While nations and factions within them dispute technical issues such as the appropriate date from which to set a benchmark for emissions and the mechanism for measuring and reporting emission reductions, the conditions continue to worsen. Those with a vested interest in continuing fossil fuel use have the economic and political power to influence actions, especially at the national level, and create conditions that make political solutions more difficult.

In addition to questions of disparities between nations separated by geography, there are also ethical issues at stake across time. Unlike environmental challenges of the past, the very fact of climate change presents ethical issues because the status quo position—doing nothing differently regarding greenhouse gas emissions—causes harm to all living things now and for future generations. Even if scientists are uncertain as to the exact quantity and exact location and nature of the harm, there is no dispute that without limiting or reversing greenhouse gas emissions, the quality of all life on earth will be degraded. For each generation, the problem of finding solutions and adaptations becomes more difficult if the parent generation has failed to take effective action. If people today do not take care to preserve the life support system of the earth for their children, they bear the responsibility for diminishing the well-being of future generations.

Climate change and global pollution present a particularly difficult set of challenges because they represent the culmination of years of actions for which the consequences either were not known or were ignored. While those who promulgated the Industrial Revolution could not foresee the environmental difficulties that would result from their actions, modern knowledge about the effects of past actions entails an obligation to prevent harm for the future. The carbon dioxide prevalent in the atmosphere today represents the cumulative combustion of fossil materials for the last hundred years. The cycling of carbon between the solid Earth and surface reservoirs in the atmosphere and oceans occurs over the course of hundreds of thousands to millions of years and more. Burning fossil fuels moves carbon into the short-term surface reservoirs far faster than natural processes can return it to the long-term rock reservoir.[25] Now, because of the long-term nature of the greenhouse gas effects, the current generation is experiencing the effects of past generations' actions as well as those of the present. The effects are cumulative for each generation forward, with the conditions becoming more acute for each generation.

The speed of the transfer of carbon from fossil sources into the atmosphere and ocean reservoirs is unprecedented in geologic history, with uncertain and unknown outcomes. Of greatest concern is the possibility that the continued increase in greenhouse gas emissions will reach a tipping point beyond which irreversible damage occurs. Thus, this generation has the obligation to take action toward stabilizing the balance of greenhouse gas emissions; each year that action is deferred makes the burden greater for the future. In the release of the 2014 Atmospheric Greenhouse Gas Bulletin, the secretary of the World Meteorological Association, Michel Jarraud, said:

> We have the knowledge and we have the tools for action to try to keep temperature increases within two degrees Centigrade (2°C) to give our planet a chance and to give our children and grandchildren a future. Pleading ignorance can no longer be an excuse for not acting. Carbon dioxide remains in the atmosphere for many hundreds of years and in the ocean for even longer. Past, present and future CO_2 emissions will have a cumulative impact on both global warming and ocean acidification. The laws of physics are non-negotiable.[26]

AN ENVIRONMENTAL ETHIC TO GUIDE A SUSTAINABLE FUTURE

> *We have barely begun to solve the problem with which cultural evolution has presented us: how to live in large groups, perpetually intensifying our activities, creating technologies few can understand and even fewer can control, without sowing the seeds of our own destruction.*
>
> **PAUL EHRLICH**

Ethics framed exclusively around human benefits have contributed to the problems of human-caused degeneration of the natural world. Now it is time to examine an earth-centered ethic, an ethic that views preserving the environment as a virtue, as guidance to move toward a more sustainable future.

Generally speaking, environmental ethics hold that destruction, overuse, or excessive appropriation of nature is morally wrong, and offers reasons for limiting human actions accordingly.[27] Over the last fifty years, specific formulations of environmental ethics have ranged over a wide spectrum of philosophies, from the Deep Ecology movement, which holds that all plants and animals have value with implications for human social issues, to anthropocentric value systems that place humans at the center of consideration.[28] Scholars of environmental virtue ethics argue that preserving the environment can be considered a virtue similar to truthfulness or integrity because preserving the natural environment is in the enlightened self-interest of humanity.[29] As noted by leading environmental ethicist Philip Cafaro, "Just as classical virtue ethics provided strong self-interested reasons for treating others with respect, so an environmental virtue ethics can provide strong grounds for environmental protection."[30]

Paul Ehrlich considers the ultimate issue in environmental ethics as defining what constitutes each individual's responsibility for maintaining crucial natural services that ecosystems supply to humanity as a whole.[31] Ehrlich's work with John Holdren established a broadly accepted relationship defining the total environmental impact of human activity, depicted below as a formula where human environmental impact (I) equals the total population (P) multiplied by affluence (A), defined as units of consumption per capita, and divided by environmental efficiency of production from technology (T).[32]

$$I=P\ (A/T)$$

The I=PA/T relationship provides an evaluative tool for the possible options to control the impact of human enterprise on the natural environment. However, it offers little guidance about how to choose among options. While agreement across this entire discipline is unlikely, preserving the planet's life support system is a universally critical value that provides a beginning point for building common ground.

Thanks largely to modern sanitary and medical technologies spread widely after World War II, the end of the twentieth century witnessed a previously unprecedented explosion in human population. The current global population of about seven billion is expected to increase to approximately nine billion by 2050, and eleven billion by the end of the century.[33] The resultant pressures of an increasing global population include the intensification of resource extraction worldwide; expansion of international systems of trade and banking; growth in centralized governments and bureaucratic management systems; and a global market dependence on high-technology improvements to productivity.[34] As the population of the globe approaches eleven billion and the economic prosperity of developing nations increases, the prospect of human consumption collectively exceeding the capacity of the natural resources of the earth presents a converging series of crises in water, food, energy, and the overall condition of the natural environment and quality of life.

Depending on cultural and economic conditions, different populations have significantly divergent total environmental impacts. Understanding the relationships among population, consumption, and technology lays a foundation for understanding the ethical choices around changing the pattern of human activities. Typical city dwellers have little idea of where food, clothing, or material goods come from as raw materials. For them, the complexity of global supply chains masks the direct relationship between consumption and resources that is obvious to indigenous people. The problem is compounded as modern urban dwellers are separated by distance from the effects of their resource use and do not perceive the feedback messages that might otherwise trigger changes in behavior.

Connecting actions to consequences in a global marketplace requires not only broad communication but empathy with our common human condition. The voices of indigenous peoples and island nations of the southern hemisphere have been especially eloquent in pleas for fairness and for a change in the direction of modern development. Here is part of the statement of the president of the Philippines at the opening of the COP21 in Paris:

> We are all aware of how the discourse on development and inequality, within and among nations, is intertwined with climate change. Invariably, those who have the least bear most of the burden. . . . For the Philippines, climate change means sorrowful catalogues of casualty and fatality; the countless voices of the homeless and the grieving—their very tears and screams carried to us by the winds and the waves that blew their homes away. During moments of great violence and bereavement, "victim" is an inadequate word to capture the loss and damage visited upon us. Each body count has a name and an age—is workmate or lover, neighbor or friend, son or daughter, father or mother.[35]

While ethicists and philosophers can argue the fine points of the differing types of ethics positions, it is essential to have a framework that is accessible in practical ways for people faced with decisions about how they relate to the natural world. This relationship is not self-evident in modern times. The capacity to feel compassion for nonhuman living things, or even other humans not in immediate proximity, has eroded in the urban, material-focused industrial societies of developed countries. Resolving the ethical challenges we face today with fairness and intergenerational justice requires a pathway toward a society more attuned to the voice of the earth.

Compassion and empathy for all the living parts of the earth runs as a common theme through the teachings of the world's major religions and spiritual practices. Their pronouncements elevate concepts such as the sanctity of all life, fairness among all people, and concern for future generations. Climate change thus poses intrinsic ethical and moral questions for diverse faith communities, based on a common concern for future generations and the viability of the natural world as the manifestation of creation. All humans also share the need for a viable planet with the nonhuman living systems upon which all life on earth depends.

Although religion and ideological differences certainly contribute to the strife and divisiveness in global affairs, the underlying threads of common concern offer hope of rising above factional disputes to address the broader underlying issues. The Papal encyclical *Laudato Si'* offers a rationale for including religion in the global response to climate change. This encyclical reverberated around the world with its detailed analysis of the human responsibility to protect the living earth, including not only people but also the nonhuman life on earth that represents the glory of creation:

> Science and religion, with their distinctive approaches to understanding reality, can enter into an intense dialogue fruitful for both. . . . Given the complexity of the ecological crisis and its multiple causes, we need to realize that the solutions will not emerge from just one way of interpreting and transforming reality. Respect must also be shown for the various cultural riches of different peoples, their art and poetry, their interior life and spirituality. If we are truly concerned to develop an ecology capable of remedying the damage we have done, no branch of the sciences and no form of wisdom can be left out, and that includes religion and the language particular to it.[36]

Many of the same concerns were reflected in an International Islamic Climate Change Symposium held in November 2014 in Istanbul, which developed an Islamic Declaration on Global Climate Change:

> Faith communities increasingly recognise that the climate crisis is also a moral crisis. The adverse impacts of climate change that we have witnessed so far, present a clear case for people of faith to examine the underlying moral causes of this phenomenon. It prompts faith communities to take action to halt the desecration of nature that leads to destruction of creation, human and otherwise. Furthermore this is an opportunity for faith communities to provide a vision, inspire others and lead the way in building a fairer, safer, cleaner world built on renewable energy—leading the way on a journey to an economic system that meets development goals and is also spiritually fulfilling. The Islamic faith community represents a significant section of the global population and certainly, can be influential in the discourse on climate change.[37]

There is much about the workings of the natural world's living system that still escapes human understanding, including how the very spark of life happens. Yet all people share a common interest in preserving the life support systems of the living earth—fresh water, clean air, fertile ground, and the biodiversity of species that constitute the global web of life, of which humans are only one part. The sense of reverence for nature as a primary element of how people relate to the natural world supports a position of taking precaution in how people use natural resources. The Zen master Thich Nhat Hanh was once asked what we need to do to save our world. He replied, "What we must do is to hear within us the sounds of the Earth crying. . . . Every Buddhist practitioner should be a protector of the environment. We have the power to decide the destiny of our planet. If we awaken to our true situation, there will be a change in our collective consciousness."[38]

THE PEOPLE'S UNIVERSAL DECLARATION FOR THE RIGHTS OF MOTHER EARTH

And this perhaps is the ultimate meaning of wilderness and its preser-

> *vation—to remind an increasingly urbanized humanity of the delicacy and vulnerability of all living species—of tree and plant, of animal and insect—with which man has to share this shrinking planet. As he learns to observe their interdependence and their fragility, their variety and their complexity, he may remember that he, too, is a part of this single web and that if he breaks down too thoroughly the biological rhythms and needs of the natural universe, he may find he has destroyed the ultimate source of his own being.*
>
> **RENÉ DUBOS**

Ultimately, an environmental ethic speaks to recognizing the voice of the earth. An articulation of this concept came clearly from a declaration created at the People's Climate Conference in Cochabamba, Bolivia, in April 2010, signed by 288 organizations from around the world, and carried to the COP21 talks with even greater support of nations, individuals, and more organizations. The preamble to that declaration states in part:

> We, the peoples and nations of Earth:
> Considering that we are all part of Mother Earth, an indivisible, living community of interrelated and interdependent beings with a common destiny;
> Gratefully acknowledging that Mother Earth is the source of life, nourishment and learning and provides everything we need to live well;
> Recognizing that the capitalist system and all forms of depredation, exploitation, abuse and contamination have caused great destruction, degradation and disruption of Mother Earth, putting life as we know it today at risk through phenomena such as climate change;. . . .
> Proclaim this Universal Declaration of the Rights of Mother Earth, and call on the General Assembly of the United Nation to adopt it.[39]

The inherent rights proclaimed in this declaration hold for all nations and all peoples a common acknowledgement of the place of humans in the natural world. The rights of Mother Earth proclaimed in this declaration from the nations most affected by global climate change and pollution but least contributing to them stand as a commonsense standard for judging actions to be taken. The basic Rights of Mother Earth, for all living things, are:

1. The right to life and to exist
2. The right to be respected
3. The right to regenerate its biocapacity and to continue its vital cycles and processes free from human disruptions
4. The right to maintain its identity and integrity as a distinct, self-regulating, and interrelated being
5. The right to water as a source of life
6. The right to clean air
7. The right to integral health
8. The right to be free from contamination, pollution, and toxic or radioactive waste
9. The right to not have its genetic structure modified or disrupted in a manner that threatens its integrity or vital and healthy functioning
10. The right to full and prompt restoration from the violation of the rights recognized in this declaration caused by human activities[40]

An ethic grounded in these fundamental rights of the Earth as a living being extends to all living things on earth. This framework can support the decision process in the face of unknowable consequences of human actions. It also places the responsibility for action squarely on each human individual, institution, corporation, nation, or assembly of nations to align with a common interest in preserving the living earth.

By preserving and protecting the ability of Earth to serve the vital ecosystem functions, all of humanity can survive and thrive on this planet. The processes that generate oxygen-rich air, filter and purify water, create food and fiber from photosynthesis, and many others, come from the interconnected operations of the living systems of the planet. Humans are but one part of that web of life. Extracting resources without any thought to replacement, restitution of disruptions, or protection to vital components has brought civilization to the brink of extinction. It is time for a just transition to a renewable and sustainable way of living based on preserving the living earth. Initiatives to a new way forward emerge in cities such as Pittsburgh with visionary leadership and communities committed to just and equitable solutions that respect the rights of all living things.

To examine the ethical challenges posed by human degradation of the natural environment, with Pittsburgh as a place-based focus, two visionaries set an ethical context for the issues addressed in the following chapters. Rachel Carson, a native of Pittsburgh, offers an eloquent vision on the value of living systems; and Karl-Henrick Robèrt, through the Natural Step Framework, offers a pragmatic, tested approach for preserving the living earth. In the keynote address at the 2010 Rachel Carson Legacy Conference in Pittsburgh, Robèrt noted the visionary influence Carson had in opening the public discussion of the precautionary principle.[41]

RACHEL CARSON AND THE PRECAUTIONARY PRINCIPLE

> *Underlying all of these problems of introducing contamination into our world is the question of moral responsibility—responsibility, not to our own generation, but to those of the future. . . . I don't believe we should wait for some dramatic demonstration before making a thorough study of the potential genetic effect of all chemicals that are widely introduced into the human environment. By the time such a discovery is made otherwise, it will be too late to eradicate them.*
>
> **RACHEL CARSON**

Finding a sustainable way forward from the current path requires a different framework for making decisions and setting priorities, one not driven simply by the potential modern technology presents but defined by compatibility with sustaining the living earth. Rachel Carson's environmental ethic provides a touchstone for this framework.[42] Inspired by Albert Schweitzer, to whom she dedicated *Silent Spring*, Carson recognized that living things have intrinsic value, even if the connection to human interests is limited to an appreciation of the beauty and complexity of nature.[43]

Carson approached the world with great reverence for life—not with the abstraction of an overarching concept, but with the deep understanding that comes from knowing life in its intimate secrets from long study.[44] She wrote of the constant of natural law: "Although man's record as a steward of the natural resources of the earth has been a discouraging one, there has long been a certain comfort in the belief that the sea, at least, was inviolate, beyond man's ability to change and to despoil. This belief, unfortunately, has proved to be naïve. . . . But the sea, though changed in a sinister way, will continue to exist;

the threat is rather to life itself."[45] Though she wrote in the early days of the discipline of ecology, Carson nevertheless captured the essential concept of protecting whole ecosystems with their complexities of food chains and material flows.[46]

It is particularly appropriate to begin with Rachel Carson in a perspective from Pittsburgh because she was a child of the Allegheny, born in Springdale, a short way from city.[47] Her environmental ethic has its foundations in her experience seeing the rural town converted to an industrial center with two coal-fired power plants and a glue factory taking the place of the cow pastures and sloping riverbanks. She came of age and studied in Pittsburgh at the Pennsylvania College for Women (now Chatham University), at the height of the industrial prowess of Pittsburgh. The plumes of black smoke from factories and houses, touted as signs of progress, affronted her sense of empathy with the creatures whose lives were obliterated through industrialization. A coal power plant with a barge docking area and a river tamed with locks and dams to allow the passage of the coal barges replaced the gently flowing, clear water riverbank she knew as a child. Seeing its transformation surely broke her heart.[48]

Though her contemporary Pittsburghers often associated Rachel Carson with the loss of industrial jobs after the Environmental Protection Agency instituted controls on emissions, her contribution to the rise of environmental consciousness has been affirmed. Carson's principles played a central role in shaping the direction Pittsburgh has taken in addressing its sustainability initiatives.[49] The City of Pittsburgh was recognized as a Biophilic City, a network of cities committed to putting nature at the heart of growth, development, maintenance, and infrastructure projects.[50]

An environmental ethic based on understanding and having empathy for the natural systems of the earth offers a clear lens through which to make decisions about taking action on climate change and global pollution.[51] Carson presented an ethical framework that places the preservation of nature, rather than human desires, at the center. Many have summarized her environmental ethic as a precautionary principle.[52] As adopted by the European Union, the precautionary principle states that where there are threats of serious or irreversible damage, lack of full scientific certainty shall not be used as a reason for postponing cost-effective measures to prevent environmental degradation.[53]

THE NATURAL STEP FRAMEWORK

A sustainable pathway forward recognizes the interrelatedness of living systems, and the necessity to think of solutions in terms of systems rather than single-shot technologies. There is no silver bullet to cure the problems we face. As ineffective responses to climate change and pollution continue to retard collective action at the global level, numerous corporations and institutions have taken action independently out of concern for the future. Many have been inspired and instructed by the Natural Step Framework, developed by Swedish physician Karl-Henrick Robèrt. Robèrt offers a framework of conditions and principles for a sustainable society that builds on the same principles Rachel Carson followed in her writing. The Natural Step Framework for Strategic Sustainable Development (Natural Step), described by Karl-Henrik Robèrt, envisions a world in which human society thrives within nature's limits.

> The problem is not that we mine and use heavy metals, or use chemicals and compounds produced by society, or disrupt natural processes. It is, rather, that our industrial and economic systems have developed so that environmentally damaging practices like these will continue to grow indefinitely, having greater and greater impacts over time. These systematic increases cannot be sustained on a finite planet . . . as long as people's basic needs go unmet, we will cut the last tree, dam the last river and pull the last fish from the sea in our efforts to survive. So we can't have environmental sustainability without social sustainability. Besides, it's no use building a world where our own well-being isn't included in our definition of success![54]

As practiced for the last twenty-five years, the Natural Step has evolved a process to improve existing conditions and to make new conditions around more sustainable practices. While this may not be the only viable approach to a sustainable pathway for the future, this process has been tested and applied in many countries, businesses, and communities.[55] Because the Natural Step is grounded in documented science principles, this approach leads to system solutions that are in harmony with natural laws.[56] The Natural Step has four conditions for human relations to the earth's natural systems:

1. *Natural systems must not be subject to increasing concentrations of materials extracted from the crust of the earth*, such as heavy metals and fossil fuels. These materials should be preserved in the earth and not released into the living systems.
2. *Natural systems must not be subject to increasing concentrations of pollutants from manmade materials* such as plastics, dioxins, pesticides, and other nonbiodegradable products. The accumulation of synthetic materials in biological systems through food chains presents challenges to natural systems and to individual organisms, including humans.
3. *Natural systems should not be subject to physical degradation* from such activities as deforestation, overharvesting, overfishing, and habitat destruction. This is an especially important principle in addressing the disparities of impact among different income levels and the ecosystems of different countries.
4. *People should not be subject to conditions that systematically undermine their ability to meet their needs*, such as unsafe working conditions and insufficient pay to support necessities. This principle addresses the issue of exporting environmentally harmful processes to places with fewer worker protections or environmental controls.

To shift the entire economic premise from one based on extraction and exploitation of natural resources to one based on regeneration, system preservation, and enhancing the ecological infrastructure requires a change in thinking more than a change in technology. Building the whole economy around principles of resource preservation, recovery, and reuse cultivates the concept of an economy based on a dynamic equilibrium, rather than one based on indefinitely expanding growth.

The process of changing course using the Natural Step process uses a four-part planning model that begins with

- *Awareness*, wherein the individual, community, organization, or company defines long-term success and develops a vision for the future.
- *Baseline data* are collected and communicated to allow all concerned to understand the starting conditions.

- *Creative solutions* are developed by back-casting from the vision of success to the current conditions.
- *Decisions about priorities* then follow the course that arrives at the success goal.[57]

Confronting the challenges of climate change and global pollution successfully will require a broader discussion in making decisions about how we use resources versus preserving resources. Understanding the consequences of continued resource exploitation can bring about a deeper consideration of what is valued, including the "priceless" value of landscape, fresh air, clean water, the diversity of species, and open space.

COMMON THEMES FOR BUILDING CONSENSUS

> *We still talk in terms of conquest. We still haven't become mature enough to think of ourselves as only a very tiny part of a vast and incredible universe. Now I truly believe that we in this generation must come to terms with nature. I think we're challenged, as mankind has never been challenged before, to prove our maturity and our mastery, not of nature, but of ourselves.*
>
> RACHEL CARSON

The public debate about how best to respond to the challenges of climate change and global pollution has not explicitly recognized the many questions of ethics and justice embedded in these issues. Uncertainty about the degree of harm, its exact location and nature of damages, and specific responsibility for the harm has created distractions and confusion. However, a consensus can be forged out of a common concern for the effects of climate change and pollution among different countries, vulnerable populations, and nonhuman living things now and in the future.

A child's view, unfettered with the minutia of public policy and uncontaminated with the conflicting interests of corporations and nations, would present a simple value proposition: choose to protect the environment and share resources fairly. Among all the discourse around climate change, global pollution, and the social problems presented by poverty, a few common principles emerge from which to craft a framework for discussion among people, and among nations.

1. *Reverence for life.* Acknowledging that humans are but one part of a complex web of life rests on the premise that all living things have the right to exist, whether they have direct benefit to humans or not. Taking precautions to preserve biodiversity and protect habitats entails actions to preserve viable conditions on earth for the sake of all living things.
2. *Intergenerational ethics.* Decisions and actions taken today determine the fate of future generations of all living creatures who have no voice at the table. The rights of nonhuman living things and the unborn of future generations to have a viable planet must be considered in today's deliberations. The failure to take timely effective action imposes a cost on the future, and limits their freedom to make their own choices.
3. *International equity.* People and ecosystems around the world have the right to be protected. Exporting pollution to areas with weak regulations is unethical. Using a disproportionate share of the world's resources but requiring an equal sharing in the cost of corrective action is also unethical. This issue presents as climate refugees, people forced from lands affected by sea level rising, or prolonged drought. It also presents as ecosystem destruction and extinction of species indigenous to those areas.
4. *Environmental justice.* The transition issues associated with moving the global economy from a fossil-fueled basis to a renewable and sustainable energy system raises numerous issues of equity and justice. The different experiences from exposure to environmental harm based on income level also raise ethical questions of responsibility for taking action. Those most in need of mitigation are often those least able to provide the resources for change. Environmental harm in the form of pollution often has more acute effects on vulnerable populations, especially children and the elderly. Workers displaced by fossil fuel curtailment have expectations of fair treatment and equity in a workforce succession.

These commonly recognized principles do not offer a prescription for ethical action. Rather, they offer a framework with which to examine the ubiquitous and urgent problems of climate change and global pollution. Finding a way forward requires action in an arena of uncertainty because, even though it is established in fact that climate change is already occurring, the future pace and trajectory of changes is not knowable. Continued scientific knowledge and innovations may offer future ways to remediate the harm, but the obligation to take action as soon as possible is tied to existing knowledge that shows the existence of harm to all living things now. Foreknowledge of harm creates the obligation to act for the sake of the nonhuman life of the planet, and to preserve more options for those in the future who are affected by today's actions, but have no voice in the decisions being made.

A Danaus plexippus plexippus, commonly known as a monarch.
JAMES MONROE

2

PRESERVING THE LIVING EARTH

Mother Earth, lately called Gaia, is no more than the commonality of organisms and the physical environment they maintain with each passing moment, an environment that will destabilize and turn lethal if the organisms are disturbed too much. A near infinity of other mother planets can be envisioned, each with its own fauna and flora, all producing physical environments uncongenial to human life. To disregard the diversity of life is to risk catapulting ourselves into an alien environment.

E. O. WILSON

WHEN RACHEL CARSON was working at the Fish and Wildlife Service in 1946, she wrote the *Conservation in Action* series of pamphlets extolling the benefits of the National Wildlife Refuges. Her first piece, on the Chincoteague National Wildlife Refuge (NWR), describes the benefits of preserving special places for those international travelers the migratory shorebirds. Chincoteague NWR, located primarily on the Virginia side of Assateague Island, consists of more than fourteen thousand acres of beach, dunes, marsh, and maritime forest. Originally established to provide habitat for migratory birds (with an emphasis on conserving greater snow geese), Chincoteague NWR today provides habitat for waterfowl, wading birds, shorebirds, and songbirds, as well as other species of wildlife and plants. Refuge staff manages this barrier island habitat to allow many species of wildlife to coexist, each establishing its own place in the environment.[1]

In August 2011 I accompanied Chincoteague NWR director Lou Hinds on a personal tour following the trail Rachel Carson traversed in 1943 on her exploratory visits for *Chincoteague: A National Wildlife Refuge*.[2] She arrived by boat traveling across the channel and through the marsh from Assateague Island to Chincoteague Island. By Jeep, well swathed in mosquito netting and with field glasses and notebook, she would have come to the sand dune reef of this barrier island. Seventy years of dune succession greet visitors now. White pines eke sustenance from the sandy soil where Carson would have seen meadows full of grazing cattle and horses. These trees stand straight and tall, holding the soil with their deep roots, supporting a lacy canopy. The sunlight that penetrates to the forest floor through their interlocking branches finds a sparse understory of a few lichen-covered live oaks supporting leafy umbrellas beneath the tall pines. Some of these trees are old enough to show the char of past controlled burning.

Further into the woods, following the sand dune ridgeline, the understory thickens with catbrier and grasses, adding texture to the succulent purslane and lichen covering the sandy ground. Around a curve in the trail appears a swale of marshy grass bounded on each side by tall trees. The grasses grow in freshwater swales running in north–south fingers among the trees, offering food and shelter from wind and weather to the ducks that come from northern continents to feed and rest, either for the winter or on the way to destinations across the Atlantic Ocean and farther south. Hinds said, "This swale signaled [to] Rachel Carson that this place is perfect for a migratory bird haven."[3] Several such swales of marsh lie within the woods—some naturally sustained by the ducks feeding in season, and some periodically mowed and flooded.

In the wildlife refuge, hundreds, even thousands, of migrating shorebirds find what all creatures need—shelter, food, security from danger. The seasonal cycles of global bird migration inspire our sense of awe and mystery. As people share these special places with transient migratory birds, for a small space of time we are coinhabitants of this fragile coastal habitat.

Chincoteague NWR is one of the most-visited refuges in the nation, providing approximately 1.4 million annual visitors with extraordinary educational and recreational opportunities. Hinds noted that more than 320 species of birds are known to occur on Chincoteague. It has been designated a globally important bird area, is part of the Western Hemisphere Shorebird Reserve Network, and has been designated as one of the top ten birding hotspots in the United States by the National Audubon Society.[4] It has a symbiotic relationship with the neighboring town, whose economy rests heavily on the visitors who come to see the ponies and visit the shore and the wildlife refuge.[5]

The tension between the town and the refuge is symbolic of the tension between natural ecosystems and the human development that too often lays waste to the wetlands and shoreline. In 1943, the Fish and Wildlife Service sequestered this area from the community that once lived here, and took the land by eminent domain to form Chincoteague NWR. People from the town still maintain hard feelings against the refuge because the farmers, fishermen, hotel, and community of residents were forced to move from the island to the mainland.[6] The two areas are now connected with a bridge, and the famous running of the Chincoteague ponies on the fourth of July attracts hundreds of people each summer. The film *Misty*, based on the children's book of the same name by Marguerite Henry, has endeared the place to millions.[7] The Chincoteague Fire Department ponies are kept to a herd of about 150 to be in balance with what the island will support, and the rest are sold for revenue.

Though not native to Chincoteague Island, the ponies are integrated into the wildlife refuge management scheme. Their droppings add nutrients to the sandy soil, and they help maintain the meadows as they crop the grasses. The ponies have adapted to eat the salt grasses and drink from the freshwater pools that form. On my tour, Lou Hinds pointed out a herd of about eighteen ponies with a large escort of cowbirds and cattle egrets, some riding on the backs of the ponies to eat the pesky flies, and some darting between their legs on the ground, retrieving the insects startled up from the grass with their passing.

Hinds said, "The service of preserving the biodiversity of our planet by offering refuge to migrating birds and wildlife enriches our life. There is more value here than the tourism and the land. It is a constant battle to keep this refuge. We fend off developers in each budget cycle, even as the sea claims more shoreline with each major storm."[8]

Rachel Carson recognized the need for preserving special places as national wildlife refuges to offer habitat and shelter for native wildlife and for migratory species as they traversed the globe on their journeys. The National Wildlife Refuges set aside and specifically manage particular lands to maintain the features that contribute value as natural wilderness. Without preserving these crucial areas from the addition of buildings, roads, and human activities that compromise the natural system, many species would perish. The human infrastructure of sea walls and buildings also constrains the formation and functions of the coastal wetlands. Natural systems have coevolved over millions of years, but the services and functions they provide have been taken for granted by people in modern times. An ethic of preserving the living earth is based on recognizing and giving a high priority to the natural systems that support life on earth. National Wildlife Refuges preserve places close to a natural state as reserves for species and as places to study the natural systems on the Earth in operation.

THE GIFTS OF THE LIVING EARTH

> *Man does not live apart from the world; he lives in the midst of a complex, dynamic interplay of physical, chemical, and biological forces, and between him and this environment are continuing, never ending interactions.*
>
> **RACHEL CARSON**

Floating on the swelling tides at the surface of the open ocean, a dense cloud of microscopic organisms pulses and throbs in the sunlight. Among them, phytoplankton filters the water through its cells, capturing the sunlight to form sugars, and releasing the oxygen into the air. These small creatures serve the earth with 50 percent of the free oxygen in our atmosphere.[9] Unseen, unnoticed, they live their cycle of life rising and falling with the tide to give breath to the earth. At the base of the oceanic food chain, they nourish creatures from zooplankton to whales.

The sun streams warm on the meadow where the air hums with bees and small beetles, flies, aphids, and ants weaving their way through the heavily pollen-laden grasses and meadow flowers. Queen Anne's lace, goldenrod, and timothy grass form a fragrant tangle over the thatch of ground. In the rich soil millions of fungi, bacteria, nematodes, and insects break down the detritus to release the embedded nutrients to the seeking roots of the plants above. Quietly, the pulse of life flows through each plant, uniting the sun and nutrients from the soil to create oxygen and sugars through photosynthesis.

The miracle of capturing light to excite molecular changes creating carbohydrates and releasing oxygen is a defining attribute of our living planet. Photosynthesis, the photochemical reaction in the living cells of terrestrial plants and phytoplankton of the oceans, makes the life of air-breathing creatures—such as people—possible. The phytoplankton in the top eighteen inches of the ocean and the forest canopies of the temperate rainforests sustain almost all of the free oxygen that air-breathing creatures depend on for life.[10]

Oxygen constitutes 21 percent of the atmosphere of the earth now, and has been stable for about ten thousand years, but it was not always so.[11] Life as it exists here and now is likely unique in the universe as its living elements shaped and changed the inert minerals and formed its oxygen-rich atmosphere. In prehistoric time, as now, the air was mostly nitrogen, but with carbon dioxide and ammonia as the other predominant atmospheric gases. The first organisms living in primordial bogs captured elemental chemicals dissolved in water welded with the energy of sunlight to form sugars, starches, and thus the foundation of life. The activity of these prehistoric organisms two to three billion years ago began to change the atmosphere by releasing oxygen. As it accumulated, the oxygen changed conditions for the emerging life forms.

The living and nonliving parts of the earth coevolved to form the complex, interconnected systems present now. Over eight million species of organisms have evolved on Earth, many not yet documented.[12] Evolution over thousands to millions of years has adapted every species on earth to the particularities of the environment in which it lives.[13] The earth has provided all of the conditions necessary to support this complex, interconnected web of life: fresh water, oxygen-rich air, fertile ground, and the biodiversity of species that share the planet. Earth's self-contained system balances the living and mineral components with elegant complexity.

ECOSYSTEM SERVICES

Ecosystems are defined by the relationship among the living and nonliving components of a geographic space, and control the flows of nutrients, energy, and organic matter through the environment.[14] Ecosystem services represent the benefits human populations derive, directly or indirectly, from natural ecosystem functions.[15] These critical life-supporting attributes of the natural earth, its diverse creatures and intricate systems, form the natural capital of the earth upon which all human enterprise depends. As nontradable public benefits available to all living creatures, most ecosystem services are not explicitly recognized within economic systems in terms of market value or economic product, even when markets depend on them for essential services.

These intrinsic attributes of the living earth differ from natural resources in that they are based on the natural interactions between the mineral and living components of the planet as they have evolved symbiotically over millions of years. While natural resources have been monetized as part of the economic system, their use by humans represents a one-way consumption without reciprocal contribution to replacement. Human exploitation of natural resources in modern times degrades the ecosystem services without recognizing the value they contribute.

Four functional categories of ecosystem services from the living earth support the economy, culture and our very existence as living creatures: supporting services, regulating services, provisioning services, and cultural/spiritual resources.[16]

Supporting services transfer materials between the living and the nonliving sectors. They include processes that define the interrelationships and flows of materials in the living earth cycles—like soil formation, nutrient cycling, photosynthesis, and fermentation—that occur as part of the life cycles of natural systems. Yeasts, as they grow and divide, consume sugars and create carbon dioxide that ferments wine and beer and leavens bread. Likewise, bacterial growth controls the formation of cheeses. Bacteria in the soils, especially rhizobia, absorb nitrogen from the air in ways that allow plants to take up nitrogen as a nutrient, a critical function in plant growth. Fungi extend plant root system effectiveness by hundreds or even thousands of times as their filaments intertwine with the finest root ends to transfer nutrients from the soil. Other organisms decompose organic material and return the nutrients to smaller, soluble elements.

Provisioning services are the processes that result in products harvested from the ecosystem, including food, fiber, and fuel; biologically derived chemicals and pharmaceuticals; fresh water; and oxygen. Unlike supporting services, these provisioning services have been most incorporated into the market system because they govern the production of goods that can be traded. The provisioning services have been exploited extensively through agriculture for food, fiber, and tobacco; timber production; commercial and farmed fishing; animal husbandry; and some pharmaceuticals.

Regulating services flow from regulating ecosystem processes from the intrinsic operation of chemistry and physics with living systems. These include the hydrology cycle of water distribution around the globe, climate regulation, or water purification, among other processes. For example, the great flow of ocean currents driven by the differences in density between fresh- and saltwater, and warmer surface water vs. colder, deeper water controls the annual cycles of the climate. The movement of the currents mixes nutrients from the bottom to the top, and from colder depths to warmer surfaces. Similarly, the control of climate happens by the flow of the temperature gradients in the oceans moving the currents in a great belt around the globe.

The hydrological cycle moves water between liquid and gaseous forms, controlling cloud cover, temperature variation over the day, and across seasons. Freshwater forms by evaporation from the surface waters and salt seas, condensation into clouds, and precipitation. Water filtering through woodland forests, wetlands, or sands is purified of contaminants as a natural process. As water flows through the ground, contaminants are removed by adhering to soil, being impeded by physical barriers, or by removal through the action of plants taking up materials, or bacteria breaking them down to elements.

On seashores, dunes and grasses form a buffer between the sea and the land. The ever-shifting shape of the shore offers a border to buffer the land from the battering waves. The seashore filters the effluent of the land as it flows back to the sea, and forms a barrier to the salt spray. The seashore has adaptive characteristics in places that are open to the flow of the waves compared to places along the shore that are bordered by walls, barriers, and buildings. When ocean storms come, human infrastructure often gives way into shattered fragments, but the shoreline swells, shifts, and reforms as the forces of water and land adjust. The resilient natural shore stands in stark contrast to the frailty and futility of human structures faced with the ocean's full force. This contrast is clearly visible in places like Chincoteague NWR where the shifting natural shoreline in the refuge adapts to storms while the towns have breaches of the sea wall, erosion of house foundations, and flooding of coastal buildings and structures.

Cultural and spiritual services are the nonmaterial contributions of ecosystems. They are less tangible, but no less vital, than the previous three. The benefits of recreation in the natural environment and the value of a natural landscape are universally recognized, and timeless. Natural cycles and the infinite variety of the organisms that inhabit the earth provide endless opportunities for inspiration to everyone, especially to artists, musicians, and writers. For the scientist, intimate glimpses of the wonders of nature, through the lens of a microscope, or field glass, for example, motivate curiosity and new challenges for discovery. The natural world offers connections to the deeper meaning of life, and allows people to assume an attitude of reflection and reverence for the elegant complexities evident in the patterns and interactions among our fellow living creatures. Even natural processes such as the formation of snowflakes or crystals in a geode offer a provocation to marvel. The sounds of birdsong and insect trilling, the movement of wind through trees or grasses, inspire the rhythms of life in music.

Understanding and preserving ecosystem services, the gifts of the earth, is the most important pathway to our sustainable future. Global human population growth and increasing affluence manifest serious threats to ecosystem services and functions: habitat fragmentation and destruction; unsustainable harvests; pollution of air, water, and land; climate change; and the spread of exotic and invasive species.[17] Every place on earth today bears the mark of human actions. Paved surfaces and buildings stretch for miles. Forests and wetlands yield to plows, saws, mines, and drill rigs. Air and water pollution penetrate to the most remote areas. Almost one-third of the world's ecosystems have been transformed or destroyed, another third heavily fragmented and disturbed, and the last third is already suffering from invasive species and pollution.[18] More seriously, the complex canopy of the Amazon forest—part of the perennial equatorial forest canopies that provide about half of the free oxygen in the atmosphere and sequester carbon in starch, sugars, and biomass through photosynthesis—has been reduced by 20 percent from 1970 to 2010.[19] Over the last 150 years, the oceans have absorbed 25 percent of the carbon dioxide from human activities, resulting in an increase in acidity of 30 percent.[20] Warmer ocean temperatures cause more evaporation, with more precipitation around the globe.

Scientists cannot predict exactly what the effects such changes will have on specific areas of the earth; however, it is clear that human actions are compromising biodiversity in both terrestrial and marine ecosystems and reducing the resilience of earth's living systems.[21] There are many choices for actions that will help to preserve and restore the functions of the living earth. Pioneering people, communities, corporations, and government entities have begun to show a new direction toward preserving essential ecosystem services.

INDIVIDUAL PATHWAYS TO RESTORING ECOSYSTEMS

> *I pledge myself to preserve and protect America's fertile soils, her mighty forests and rivers, her wildlife and minerals, for on these her greatness was established and her strength depends.*
>
> **RACHEL CARSON**

From the early days of the Industrial Revolution, union workers clamored for safer working conditions in the mills, characterized by brutally long hours and dangerous working conditions including lack of ventilation, no protection from machinery, and the constant roar of machines and fires.[22] Worker safety was especially ignored in the vigorous economic activity surrounding the production of war munitions and materials for both World War I in 1914 and even more intensely in the World War II period. In Pittsburgh, worker productivity was paramount in places such as the Westinghouse Electric, Machine and Meter plants where the management practices of Fredrick Taylor aggressively implemented incentive pay, time studies, and piecework systems.[23] On June 3, 1914, in an unprecedented heat wave, twenty-three men walked out of the plant protesting poor ventilation in the shops filled with mica dust. Workers complained that the air was so bad, they were sick more than half the time because of their working conditions. People endured the smoke, heat, and noise of the mills for the sake of their wages, and for the hope of a brighter future for their children. The exploitation of the workers, almost to the point of slavery, lies at the foundation of Pittsburgh's industrial might.

The struggle between workers' rights and the corporations that provided the investment capital played out in an epic drama over fifty years.[24] Eventually, much of the heavy polluting industry moved overseas, where the demands for justice and the laws protecting the environment were either absent or less enforced. Lawrence Summers, chief economist of the World Bank in 1991, proclaimed the benefits of international trade liberalization thus: "I think the economic logic behind dumping a load of toxic waste in the lowest-wage country is impeccable and we should face up to that. I've always thought that the countries in Africa are vastly under-polluted; their air quality is probably vastly inefficiently low compared to Los Angeles. Just between you and me, shouldn't the World Bank be encouraging more migration of the dirty industries to the Least Developed Countries?"[25]

The ethical implication of this kind of thinking extends the concept of the interconnectedness of global life as well as global economics. What Rob Nixon of Harvard describes as the "slow violence" of environmental pollution presents an environmental justice issue all over the world.[26] Is it acceptable to cause health-threatening pollution far away when it is not acceptable close to home? If modern-day consumers knew the effects of this behavior, would they have a different attitude toward products they purchase?

Individuals make thousands of decisions each day that shape how they connect to the earth. Some patterns can change quickly, and some need longer periods of adjustment and response. Every person can make a difference in the way he or she uses the resources of the natural world, as each person shapes a lifestyle aligned with his or her personal values. Collective personal actions shape the momentum of society.

PERSONAL CONSUMPTION: NEEDS OR EXCESS?

The average American generates about 1,598 pounds of trash each year, and only about 37 percent of the discarded material is recycled.[27] Individuals can make a huge difference in the health of ecosystems by reducing waste, and plastic waste in particular. Scientists estimate that every square mile of ocean contains forty-six thousand pieces of plastic.[28] The amount of plastic accumulated in the ocean since the end of World War II is estimated at three hundred trillion pounds, greater that the total estimated weight of all the fish in the sea.[29] This material is accumulating in gyres miles wide and hundreds of feet thick, impeding the life of plankton, and carrying toxins and bacteria through the ocean food chains.[30]

As part of a class in geology and environmental studies I taught at the University of Pittsburgh, a group of students tracked everything they used in one day, identified the origin and mapped where all these things come from, and where they ultimately end up. A global supply chain emerged: Pittsburgh students use resources from all over the world, and most of the goods end up as trash.[31]

Several questions arose from this exercise: Do we actually *need* all of the kinds of things we have? Can we consciously reduce the amount of clothing we have by sharing with less-fortunate people, or by trading with friends instead of buying new? Can we develop more meaningful ways of showing respect and affection than by buying more material things? Can we value quality and long life of goods rather than short-lived fashions or trends? Does anybody actually need air freshener?

The students were inspired by the adventures of filmmaker Mark Dixon, who came to the class and showed his film *Your Environmental Road Trip* (YERT). As part of the Rachel Carson Centenary celebration in 2007, Dixon and his friends Ben Evans and Julie Dingman-Evans set out from the Rachel Carson Homestead in Springdale, Pennsylvania, to take a yearlong tour of all fifty states. Their aim, on the trip, was "to document environmental sustainability in every state in the union in a single year. We wanted to find out: What can Americans do to save the planet? What are they already doing? Why aren't we all doing more?" In the process, they "conducted over 800 interviews, amassing nearly 600 hours of footage." However, the filmmakers were interested not only in documenting others' efforts but in attempting to live the principles they were exploring. In their words:

> We were on a mission to personalize sustainability, and as the National Academy of Sciences pointed out back in 1990, humanity is currently "conducting an uncontrolled experiment with the planet." So as we traveled and filmed, we decided to BE the experiment—we would never turn on an incandescent light bulb, we would radically reduce our water usage by any means necessary, and, above all, we would attempt to generate zero waste, keeping all of our garbage (and recyclables) with us for the entire year . . . in our packed-to-the-gills hybrid. To pull this off, all of our garbage each month would have to fit in a shoebox.[32]

The YERT Project helped to raise students' awareness of the supply chain of everyday things that depend on resource destruction or extraction from far away. As this project demonstrated, individual households can set a goal to reduce the amount of material discarded each year by half. The team demonstrated how to manage this by asking a series of questions at the point of purchase: Do I need this? What will I do with it when I am finished with it? Can I recycle, reuse, or trade it? Can I compost it? Is there a domestic, local, or biodegradable alternative? Is there a less toxic alternative?

Every person can reduce waste. The practice of choosing less processed and packaged goods at the point of purchase can cut down on excess plastic. Most communities have a recycling program making it easy for people to recycle glass, plastic, and cans—sometimes also cardboard and paper. Two actions everyone can take would reduce the plastic burden significantly: one, carry a reusable bag for incidental purchases and larger bags for groceries and bulk purchases; and two, cut plastic bottles out of your life to the extent possible. If you must use them, recycle them diligently. If you want to really make a difference, pick up the incidental plastic trash that gathers in gutters and along edges of streets. Much of this carelessly discarded plastic waste ends up washing through storm sewers and finds its way into bodies of water where aquatic creatures become entangled, starved, or poisoned.

CULTIVATE A LIVING HABITAT IN URBAN AND SUBURBAN NEIGHBORHOODS

Since the end of World War II, the backyard has been a part of the American dream accompanying home ownership. The suburban cliché of a green lawn with one tree and a picket fence is often maintained by using herbicides, pesticides, and fertilizer in alarming amounts.[33] These chemicals may indeed guarantee a "weed-free" lawn and allow only selected flowering species to grow, but this practice imposes a cost to the environment in the form of water contamination, and suppression of native birds, pollinators, and creatures of the living soil.

Backyards can increase habitat for wildlife, grow some food, and help to control stormwater runoff. The greatest improvement comes from eliminating or sharply reducing the use of pesticides, herbicides, and chemical fertilizers in favor of natural landscapes and chemical-free cultivation. Preserving habitat to encourage biodiversity requires little effort, and brings great rewards. A backyard with enough open cropped meadow of violets, white clover, and creeping jenny to encourage ball tossing, tumbling, or play can be surrounded by borders of hedges that bear fruits and flowers for birds and pollinators. Judicious selection and spacing can provide year-round beauty, food and cover for birds and small animals—sometimes even bigger animals.

In western Pennsylvania, a hedge alternating mountain laurel, forsythia, rose of Sharon, and privet hedge gives a delightful variety of texture. Rain gardens and rain barrels at the corners of the house can help to water gardens and control the flow of stormwater runoff as well. Trees in the center corners of the yard, away from power lines and the street, can provide shade and shelter. If neighbors join together to plant habitat corridors, a wildlife refuge in miniature can flow through the neighborhood. The benefits of tempering the climate with shade and windbreak as well as supporting a diversity of birds and wildlife add enjoyment and health to the local personal environment. Simple solutions work. Planting an assortment of milkweed, star aster, goldenrod, black-eyed Susan, and pennyroyal creates a garden that requires little care and returns every year providing food for bees, birds, and butterflies.

Habitat that encourages biodiversity and ecosystem services works for common space in apartment buildings or spaces between buildings in city lots. Using native plants for wildflower gardens can work in all parts of the country.[34] The City of Pittsburgh converted from pesticide and herbicide spraying for right-of-way maintenance to integrated pest management practices in 2007.[35] The cost savings for reduced chemical purchase more than covered the cost of training and labor for maintenance workers. The random spaces between paved surfaces in Pittsburgh now often bloom with seasonal wildflowers.

PITTSBURGH PATHWAYS TO PRESERVING NATURAL SYSTEMS

For all the people, the preservation of wildlife habitat means also the preservation of the basic resources of the earth, which men, as well as animals, must have in order to live. Wildlife, water, forests, grasslands—all are parts of man's essential environment; the conservation and effective use of one is impossible except as the others are also conserved.

RACHEL CARSON

Pittsburgh offers an excellent stage upon which to see the challenge of restoring ecosystem functions. As the industrial character of this area yields to the biomedical, electronic, and high-tech economy, the well-known legacy of the industrial past lingers. The battle for clean air has been a major part of the history of this city. The early battles against the heavy, sulfurous smoke from the steel mills and coke ovens that once rolled through the valleys continue today against the fine particulate matter and unseen contaminants like ozone and benzene. Air quality remains a challenge from existing power plants and industrial activity, transportation emissions from rivers, roads, and rails, and air pollution from neighboring places. Aging infrastructure challenges the capacity of water treatment facilities. Combined storm and sewer systems overflow contaminants into the rivers. Hard paved surfaces cover much of the land, in places scarred with abandoned and crumbling buildings or empty industrial sites.

Citizens of Pittsburgh have chosen to accept the challenge of stopping these negative effects, although the struggle with past patterns continues to play out in conflicts among various interests today. For example, the boom in Marcellus Shale development introduced new enthusiasm for manufacturing, especially of synthetic plastics and chemicals based on accessible, relatively inexpensive natural gas. In counterpoint, champions of ecosystem restoration argue for initiatives to build a more resilient and sustainable life of this city. An ethic of preserving the living earth is emerging. New initiatives have sprung up based on the increased value of a quality of life that includes clean air and water, and environmental amenities such as parks and recreation opportunities.[36] Here, in the area Rachel Carson called home in her formative years, the restoration to a more resilient landscape to support ecosystem functions is becoming a clear and growing priority. In addition to opposing Marcellus Shale development from concerns about water quality impact, active citizens and business organizations propose green infrastructure for stormwater management. Today, ecosystem services are improving in Pittsburgh.

The cases presented here show actions taken from an ethical perspective, rather than for purely economic reasons, to restore, preserve, or regenerate essential ecosystem functions and services.

PITTSBURGH BOTANIC GARDEN: AN EXERCISE IN RECLAIMING A COAL MINE

A rolling hillside covered with several inches of spring snowfall reveals the contours of land once home to the Iroquois. The Allegheny County Settler's Cabin Park was a retreat close to the city where I would go often during my years as a graduate student to escape into the relative peace of the woods. Parts of the park overlooked the scars from the strip mines that had operated there in the 1940s, and three of the four streams in the area ran orange with acid mine drainage. This land, colonized by mostly Scots-Irish and German settlers, was used for farming and raising livestock in the 1700s, but cultivation was difficult from the steep slopes and the shale.[37] According to the website of the Pittsburgh Botanic Garden, the current occupants, "The area was deep mined in the 1920s and then surface mined in the 1940s. William J. McGill's family farmed this land for three generations. Descendant Beth McGill Ellis grew up on the family farm and recalls that after the mining, the water was not potable due to the acid mine drainage, and they had to drive into the community of Oakdale to fill jugs with drinking water. The McGill farm was sold to the county in 1971."[38] Settler's Cabin Park was carved out of part of this land, and the county initially sold part to a developer interested in making it into a golf course, but these plans did not materialize. The land sat unused and unoccupied until 1988, when it was taken over by a group of architects and horticulturists who envisioned a world-class outdoor garden for the Pittsburgh area.[39]

The successful reclamation of this land left in ruins for decades is a testimonial to the power of a vision and the persistence of dreamers who charted a different fate for this wreckage of industrial exploitation. Beginning in 1989, the Botanic Garden of Western Pennsylvania went through the process of securing a ninety-nine-year lease from Allegheny County, set about the planning and organizing to begin restoring the land, and in 2002 hired its first executive director, Greg Nace. The flood of the mine area by Hurricane Ivan in 2004 resulted in a major change in the planned approach from treating the acid mine drainage to complete reclamation of the mined areas, a more permanent solution.[40] The remaining coal needed to be removed by "daylighting" the mines, a reclamation process that "eliminates the acid mine drainage that brews in the open spaces of abandoned coal mines" by "removing the soil on top of the old mines, scooping out the remaining coal . . . and then replacing the soil."[41] To pay for the remediation, the organization has been selling coal from 142 undermined acres, as about 27 percent of the coal on the site was still in the ground.[42]

On August 1, 2014, I joined dignitaries, supporters, and volunteers for the ceremonial opening of the first sixty acres of the Pittsburgh Botanic Garden. As we walked the half mile to the Lotus Pond, we passed through fields of grasses and wooded areas with education nodes, and saw newly planted trees staked and shielded from deer. As I stood on the footbridge overlooking the pond, I watched butterflies float on the humid air and saw a blue heron rise from the water's edge. In the shallow water there, a snake slowly swallowed a hapless frog to the fascinated stares of two young boys. From the podium, Nace explained the years of work that had brought this pond to its current state of life.

The water filling the pond is piped from the acid mine drainage and passed through a vault filled with limestone. The water's contact with the limestone raises its pH—neutralizing the acidity—and lowers the concentration of aluminum, thus restoring it to a state that can support life.[43] The system, designed by Bob Hedin, won a Governor's Award for Environmental Excellence. "Prior to this project, the lifeless pond was nothing more than an unsightly liability," said Eric E. Cavazza, director of the state Department of Environmental Protection's Bureau of Abandoned Mine Reclamation. He added, "The system has restored water quality in both the stream and the pond, and the Lotus Pond now supports a healthy aquatic community."[44]

The time, effort, and expense for completing this reclamation project illustrate the burden coal extraction left as its legacy. Over three thousand miles of Pennsylvania streams and rivers are polluted with acid mine drainage from abandoned coal mines,[45] and many of the thousands of acres affected by coal mining will never receive the investment of time, money, and dedication that Nace brought to the Pittsburgh Botanic Garden. This project represents the sustained partnership among the state and federal government agencies and private industry sponsors who provided funding, and the thousands of hours of dedicated volunteer service from the community and the university faculty and students who worked on this project over a decade. Once lost, restoring a living and functional ecosystem is a challenging and expensive undertaking. But seeing healthy ecosystems restored to this once fertile land offers hope for the future.

THE NINE MILE RUN WATERSHED: RECOVERING URBAN WETLANDS

At the height of the steel industry days in Pittsburgh, sparks flew high into the air in the evenings when Duquesne Slag Products dumped the slag over the hillside. For fifty years, from 1922 to 1972, the slag piled into the valley of an urban stream called Nine Mile Run as it flowed into the Monongahela River, eventually covering 238 acres, to a depth of 120 feet in places. When the last slag was dumped in 1972, about seventeen million cubic yards filled the Nine Mile Run Valley from end to end.[46] By the end of 1970, the river as it flowed along the edge of Frick Park was nicknamed "Stink Creek" and children were warned to stay away. Indeed, all of the rivers in Pittsburgh at that time were places of dirt, danger, and pollution—industrial waterways that carried barged materials of commerce and were saturated with sewer overflow and industrial

The Nine Mile Run was once called Stink Creek due to industrial wastes and sewage overflow.

JOHN MOYER

wastes such as phenol. The decline of the steel manufacturing industry came in the 1980s, with the virtual collapse of major manufacturing, and as major mills fell quiet, Pittsburgh began a period of adjustment, diversification, and rebirth that continues to this day.

By 1998 the slag pile attracted interest as a development site for an upscale residential community, now called Summerset at Frick Park. But, as the prospective developers looked over the location, it was clear that grading and covering the slag pile would need to occur, and the polluted Nine Mile Run would need to be addressed. The simplest solution, at that time considered the only option, proposed to contain the stinking creek in a culvert and pipe it to dump into the Monongahela River further away from the development site. At this point, the miracle of Pittsburgh comes in to play. This plan was questioned by a group of faculty from Carnegie Mellon University who secured funding to evaluate an alternative plan that would restore Nine Mile Run.[47]

Among those bringing their expertise to the project was Carnegie Mellon professor Bob Bingham, who was teaching a class as part of the Studio for Creative Inquiry.[48] Students worked on the Nine Mile Run Greenway Project from 1996 to 1999. The student projects focused on the intersection between community, art, and the environment, and addressed the need for urban residents to reconnect with and positively impact the natural world.[49] The idea for a Watershed Association grew out of the enthusiasm generated by this project and the desire to continue these efforts. In the fall of 1998, the final report, "Ample Opportunity: A Community Dialogue," revealed the option to restore and recover the wetland functions of the stream.[50] Under the leadership of community people, with data collection and substantiation by university students, the cost of this option looked competitive.

The Urban Redevelopment Authority of Pittsburgh owned the property and had jurisdiction to seek and provide funds as a brownfield recovery site. Under a consent order from the Pennsylvania Department of Environmental Protection, resources were available to conduct a brownfield remediation and to correct stormwater overflow violations. The situation was pretty bad. Seeping through the slag pile gave the water a pH of about 11, more alkaline than bleach. There were few fish or other living organisms in the water.[51] Remediation included installing a French drain at the base of the slag pile to collect the heavily contaminated water and redirect it to the water treatment plant at

Numerous fish swim in the now restored Nine Mile Run. An effort that brought together local, state, and federal governments, as well as students, individuals, and nonprofit organizations.
JOHN MOYER

Allegheny County Sanitary Authority (ALCOSAN) by pipe.[52] Then the slag was graded, covered with topsoil, and planted with native plants and grasses such as winter rye.

The project was funded in part by a congressional appropriation of five million dollars under provisions of the Clean Water Act and was overseen by the US Army Corps of Engineers.[53] The City of Pittsburgh contributed some funds and in-kind services for a total project cost of $7.7 million over six years.[54] The restoration extended 2.2 miles from Fern Hollow down to Nine Mile Run and created several wetlands, removed invasive species, planted native plants, and restored bends and riffles in the stream trajectory. The broad coalition of community participants engaged in this restoration was a critical part of its success.[55] The Nine Mile Run Watershed Association, formed as a citizen awareness organization, continues to this day to monitor progress, provide maintenance through volunteer efforts, and increase community awareness and engagement throughout the watershed.[56]

As people in the upper watershed areas that drain into the rivers have become involved, they have installed rain barrels, rain gardens, and permeable paving to help control stormwater runoff. Brenda Smith, executive director of Nine Mile Run Watershed Association, said, "There are now so many fish of healthy and diverse types that counting all of them at any one point has become nearly impossible."[57] The Stink Creek of the smoky Pittsburgh past is now the Nine Mile Run watershed and wetland, a recreational and biological asset to the city, and part of the pathway to a sustainable future.

FOREST HILLS BOROUGH: AN AMERICAN TREE CITY

The Borough of Forest Hills, Pennsylvania, covers an area of about 1.5 square miles, located approximately seven miles east of downtown Pittsburgh. Founded on July 29, 1919, as part of the Westinghouse workers' housing associated with the plants in Turtle Creek, Forest Hills now has well-established, family-friendly neighborhoods, beautiful tree-lined streets, and transportation corridors connecting to the city and surrounding areas. Since its inception, Forest Hills Borough has sought to maintain its natural assets, dedicating almost one-quarter of the land area as public parks. This commitment to green spaces has been recognized by the National Arbor Day Foundation, which has designated Forest Hills as a Tree City USA community for over twenty years.[58]

The Borough of Forest Hills has a Tree and Shrub Commission that oversees the planting and care of community trees and shrubs on streets and neighborhoods. Information is provided to citizens through two active garden clubs who plant and maintain gardens in the park, around municipal buildings, and in the long corridor of Ardmore Boulevard that runs through the middle of the town. Although the steep hills are somewhat of an impediment to pedestrians, many do walk through the neighborhoods on tree-shaded sidewalks, and the two largest parks offer trails and recreational opportunities in all seasons. There are stairways called "crazy paths" through the neighborhoods that cross the hairpin turns of the road, legacies from the days when people rarely had cars but walked down the hill to the streetcar line that once ran through town. Now they provide walkable connections through neighborhoods. Where terraced farms once covered the hills, now the paths reveal shady gardens and backyards. Urban agriculture is encouraged by an ordinance specifying conditions for people to have beehives and chickens in their property. The weekly Forest Hills Farmers Market, hosted by the Late Bloomers Garden Club, provides local and organic produce from May to October and promotes contributions to a local food bank every month.

A deliberate effort to maintain a natural environment depends on the participation of the whole community. People in Forest Hills treasure their trees, and a licensed arborist assists in the maintenance and care of the street trees. The value of this approach is reflected in the relatively well-managed storm runoff water. Among the neighboring towns, Forest Hills has one of the lowest levels of stormwater runoff overflow issues, partly because of the heavily vegetated hillsides. In some areas where clearing has been done for commercial or newer developments, stormwater management becomes more challenging. The older neighborhoods where the trees rise over the houses give models for success.

Living in this community offers numerous opportunities to engage with nature in daily life, which can occur in any area where the natural environment is protected as part of the land use. In this community where people value the nature around them, each season enriches the experience of daily life—an example of a sustained ecosystem providing cultural ecosystem services. It is impossible to live in such a place without developing an acute sense of connection to the trees, birds, and other wildlife that share this space. People become attuned to the diversity and the beauty of nature as a part of their normal life experience.

CONCLUSION

As the challenges of global pollution and climate change become more acute, recognizing the services the living earth provides through its intrinsic functions becomes more important. Basic life-support systems have been endangered by the pressing short-term priorities of infrastructure, housing, agriculture, and material consumption. But as the explicit value of ecosystem functions become evident, actions to restore intact living systems receive more attention. For example, although methods of computing value vary greatly, wild pollinators in the United States contribute at least $19 billion annually in services to agriculture.[59] Recognizing the value of wild pollinators has increased interest in addressing their population declines from aerial spraying of pesticides, habitat destruction, and warming temperatures. Even when severe ecosystem disruptions have occurred, it is possible to restore some functions such as wetlands for water purification. However, the wider global effects from climate change and global pollution will require a stronger effort to prevent destruction of living systems, or to slow and reverse the severe imbalances from moving carbon from fossil deposits to the more dynamic carbon reservoirs of the air and sea. Recognizing the value of ecosystem services is an essential part of an ethic of preservation and restoration for the living earth.

The author's grandchildren Quinn and Lia inspect a sand crab at Chincoteague Island.

PATRICIA DEMARCO

3

INSPIRING A SENSE OF WONDER

If I had influence with the good fairy who is supposed to preside over the christening of all children, I should ask that her gift to each child in the world be a sense of wonder so indestructible that it would last throughout life, as an unfailing antidote against the boredom and disenchantment of later years, the sterile preoccupation with things that are artificial, the alienation from the sources of our strength.

RACHEL CARSON

A MEADOW OF wildflowers sloping toward a stream edged with woods hums with life. To some, this serene vista invites exploration and wonder at the myriad living things revealed; to others, it represents building plots and a site for development. Since the Industrial Revolution, technology has accelerated the human imprint on the natural world reshaping the landscape for roads, dwellings, cities, and industries. Preserving natural spaces, wildlife refuges, and protecting endangered species becomes more and more difficult as the pressure for human needs and wants increases.[1]

With the development of manmade physical infrastructure and concentration into cities, people have become disconnected from the natural environment.[2] Many people feel so dependent on the built environment they do not think of themselves as part of the natural world. Eighty percent of North

Americans live in towns and cities and spend 90 percent of their time indoors.[3] In the United States, 82 percent of children live in urban areas and most spend two to four hours per day using electronic media compared to spending an average of one hour per day in the outdoors.[4] Families may have little opportunity to share time outdoors where adults and children may share experiences in nature.[5]

This disconnection has accelerated the destruction of natural systems and makes actions to redress the harm more difficult. This indifference to nature is evident in broadly applied destructive technology. People who do not feel connected to the natural world see little compulsion to protect or preserve habitat or natural systems.[6] They also have difficulty in feeling empathy for other people in other places. Yet people are not separate from the natural environment; they are an integral part of it, even if they are unaware of the connection. Preserving and protecting the environment is part of survival for humans and other species. Conversely, harming the environment harms the well-being of all people, directly and indirectly.

Naturalist E. O. Wilson observed, "Nature is part of us, as we are part of nature."[7] Psychological connections with nature have the capacity to facilitate sustainable attitudes, and may be an important tool in preserving the environment.[8] While only 16 percent of American adults place preserving the environment as a high priority, especially if they think that doing so will reduce economic gains, most (49 to 52 percent) would support environmental improvements if they think it will improve their own health by improving water or air quality.[9] Connecting people to the natural environment improves health, sharpens the mind, and opens them to empathy with all living things. This sense of self as a part of the living earth is critical to preserving humanity, spirituality, and the capacity to survive as a species for millennia. An emotional connection with nature provides the context for human existence and the wisdom to evaluate what needs to be done for the sake of other living things and for future generations.

THE CASE FOR CONNECTING WITH NATURE

I believe natural beauty has a necessary place in the spiritual development of any individual or any society. I believe that whenever we de-

stroy beauty, or substitute something man-made and artificial for a natural feature of the earth, we have retarded some part of man's spiritual growth.

RACHEL CARSON

Being connected to nature offers many significant benefits to individuals and to society.

Connecting to nature promotes empathy with other people and living things. As diverse members of the same species, all people are more alike as humans than different in racial and cultural attributes. Comparing the human genome to analysis of genomes of other organisms illustrates, significantly, that humans

Stella releases an American song sparrow banded for the National Aviary Neighborhood Netwatch program.

ANNE AVONDOLIO

share over 90 percent of their genes with other animals from crocodiles to nematodes.[10] A sense of connection to the natural world creates a sense of responsibility and concern for other living things. For example, a person may develop empathy if he or she has a chance to watch robins feeding nestlings and sees the young birds fledge and grow under the care of the parent birds. In addition to regular contact with nature, one of the best ways to foster empathy during early childhood is to cultivate children's relationships with animals. Young children feel a natural kinship with and are drawn to animals, especially baby animals.[11] Care for small animals as pets often helps children develop a bond with living creatures, as does the experience of planting and cultivating flowers or vegetable plants. Children learn a sense of responsibility and pride in the success of their actions when they can harvest flowers or food they have helped to cultivate.

Connecting to nature promotes healthy minds and cognitive skills. Nature experienced vicariously is not real, especially to children, and does not create the expanded sense of self as part of the natural world. It is especially important for children to experience nature as part of their everyday world. Children who have direct outdoor experiences develop better cognitive skills, concentration, and show better self-discipline. The greener the environment, the better their test scores in school.[12] Playing outside also advances motor skills and fitness. Children who play outside in a natural environment develop better balance and coordination and experience less sickness than children who play in paved playgrounds or indoor environments.[13] When they play outside, children develop a wider imagination with more diverse and creative play that stimulates language and collaborative skills.[14]

Moreover, children who play together outdoors have more positive feelings about each other and are less prone to bullying.[15] Pretend games and building their own play places and rules builds collaboration and a sense of community. Connecting to nature helps children deal with adversity and allows an escape into nature as a way to relieve stress.[16] All of the experiences from being in a natural environment help children become more observant and creative by providing inspiration for creative expression in art, music, and writing.[17] Children become more independent and self-confident from playing in an unstructured natural environment. The critical period of middle childhood, ages six to twelve, appears to be an especially important time when playing in nature helps in developing the capacities for creativity, problem solving, and emotional and intellectual development.[18] For many children, asphalt schoolyards with fixed equipment and sports fields, exhibiting no shade trees, plants, or natural spaces are the common experience for "recess." This condition may have been less of a deficit when students went home to play outside in gardens, alleys, or local parks. But as the time for unstructured play outside has diminished steadily over the last three decades, natural places have become less and less a part of children's experience of life; thus, the children carry no strong connections to the natural world into adulthood.[19]

Connecting with nature promotes healthy bodies. To be healthy, people need a healthy environment. Fresh air, clean water, and food free from microbial, mineral, and chemical contaminants are essential for health, of course, but experiencing nature in daily life also confers benefits that have been measured objectively by examining data using the amount of green space present as a variable for metrics such as blood pressure, performance on standardized neurocognitive tests, and physiological measures of immune system functioning.[20] All of these measures of physical health were better in areas with more green space, regardless of the level of poverty and other parameters such as age.

Another objective measure of the benefit of a natural environment comes from studies of hospital patients. Surgical patients whose rooms offer views of nature have been shown to require less pain medication and have more favorable postoperative recovery experiences.[21] A study of the electronic medical records of 345,000 people evaluated the most prevalent medical conditions related to the amount of green space within one kilometer of the patient's address based on satellite images. Out of the twenty-four disease clusters examined, fifteen were significantly less frequent in greener environments, and none were more frequent.[22] In sum, a whole host of different diseases is less prevalent in green environments, and this relationship is not due to income or social status differences in the people who live in greener and less green environments.[23] While it may seem obvious that having a natural environment contributes to a person's sense of well-being and offers peace of mind and a source of emotional serenity, physical benefits also are associated with experiencing nature. For example, obesity and diabetes are less prevalent among people who have access to green spaces.[24]

Connecting to nature engenders a greater feeling of happiness and satisfaction. A theoretical basis for the notion that contact with nature is beneficial comes from E. O. Wilson, who introduced the term *biophilia*, defined as the innately emotional affiliation of human beings to other living organisms, almost twenty years ago.[25] Being in nature creates a sense of empathy with life, a sense of vitality and energized mental acuity that is not achieved from either exercise or recreational activity in a manmade space. Research designed to determine whether the association between nature relatedness and happiness is due to a general sense of connectedness or a more specific link with nature has revealed that, among the various happiness scales tested, association with nature was highly significant.[26] People who connect to nature with as little as twenty minutes per day are more content, more emotionally secure, have a more positive outlook on life, and have a greater sense of vitality.[27] In the words of Rachel Carson, "Those who contemplate the beauty of the earth find reserves of strength that will endure as long as life lasts."[28] Physician David Cumes has described "wilderness rapture," including self-awareness; feelings of awe, wonder, and humility; a sense of comfort in and connection to nature; increased appreciation of others; and a feeling of renewal and vigor. He is convinced from years of study with Bushmen of Africa that "nature's medicine is the most powerful tool of all."[29]

INDIVIDUAL PATHWAYS TO CONNECTING WITH NATURE: OBSERVATION, APPRECIATION, AND CURIOSITY

It is time for our culture to walk in the forest again. Once we see nature as a mentor, our relationship with the living world changes. Gratitude tempers greed. . . . We realize that the only way to keep learning from nature is to safeguard naturalness, the wellspring of good ideas.

JANINE BENYUS

In the summer of 2010 a group of students from the Pittsburgh Student Conservation Association gathered on the outdoor amphitheater benches of the Rachel Carson Homestead in Springdale, Pennsylvania, for the first day of "Bug Camp."[30] They had come to the two-week summer day camp to learn about the natural environment and experience time in the outdoors. None of the twelve students had a backyard, garden, or park in their own neighborhood. They made faces and groaned at the request to turn off all electronic devices for the duration of the class. They received directions and a worksheet to guide their exploration of the quarter-mile trail through the woods and meadow, with instruction on avoiding poison ivy. In response to the routine inquiry, "Any questions before we set out?" one student asked whether there would be bears or tigers, and he was genuinely concerned about safety in the deep, dark woods. Others chimed in about fears of entering what seemed to them a strange, hostile place. But by the end of Bug Camp, all of them were comfortable walking in the woods, could identify most of the birds, animals, and insects they found there, and could name the trees and many of the plants. They had witnessed the emergence of an adult cicada, watched a butterfly extend its proboscis into florets of a zinnia, mapped the path of a honeybee through a patch of clover, and had seen a red-tailed hawk swoop up a chipmunk. The songs and poetry they wrote at the end of the session captured their newfound joy in nature: "Butterfly and Firefly Waltz," "The Bad Bug Blues," and "The Rap of Life."[31] They went back to their community with seedlings to plant and a better understanding of living in harmony with nature.

Today's children too often grow up isolated from the tactile experience of exploring the natural world. Parents who feel alienated themselves may be intimidated by the prospect of teaching their children about nature.[32] In reality, parents can learn much from their children. A young child just beginning to notice his or her surroundings has unbounded curiosity and interest in even the smallest things. If this curious mind is let loose to find out about the living world of creatures and growing plants, the fundamental connection to the natural habitat of humans will form. The world needs better understanding of the rules of nature that support life, yet children are too often isolated and "protected" from the actual experience of encountering nature. Fears of allergies, violent intruders, or dirt keep children inside.[33] Viewing elephants, sharks, and lions on TV can seem more interesting and dramatic than watching squirrels, bugs, or birds in the backyard, but learning about the creatures that share one's living space offers many more important lessons than the passive observation of dramatized natural history. Computer games and electronic media offer no smells of nature, no textures and feelings, no tastes, no layered sensory stimuli compared to the experience of lying in a meadow on a warm summer day. The practice of connecting to nature can occur in any area where the natural environment is protected as part of the land use, and can be encouraged through *observation*, *curiosity*, and *appreciation*.

Observation. The first essential pathway to living in harmony with nature is to notice and value natural places everywhere. Regardless of location, everyone can walk and notice the features of the changing seasons. People can become attuned to the diversity of life in ordinary places by practicing the discipline and delight of noticing details and sharpening the powers of observation, as with a child. Simple pleasures emerge, such as noticing all the different leaves in the neighborhood and figuring out which tree they came from. Even in concrete-clad urban areas, the dandelions crack the sidewalks, vines climb fences and walls, and city birds pick up crumbs. Noticing and observing is the first step in cultivating a connection to the natural world.

Curiosity. Developing curiosity about other living things stimulates a deeper level of exploration. The curious observer can discover how the creatures in the surrounding neighborhood interact with people's lives. Where do our needs coincide? Where do they clash? What is our role as people who share space with other living things? Children who enjoy encouragement in this direction will naturally be pulled into science by their own curiosity. Many a great scientist gained entry to the field through observation and exploring. The life of renowned scientist E. O. Wilson, as well as the life story of Rachel Carson, illustrate gloriously how childhood curiosity about the natural world can flourish into a life of scientific pursuit.[34] Here, having access to a library, even online resources, can enlighten and inform, as an older child can investigate the life cycle of an insect or animal or see what the habitat or food supply consists of. A mobile app can identify the name of a bird heard calling through the trees in the neighborhood.

Appreciation of nature with all five senses. Awareness of the natural world helps to refine and expand the ordinary experiences of daily life through all five senses. Computer games and screens portray scenes with sounds and special effects, sometimes from distant, exotic places, and sometimes fabricated and surreal experiences that are not of the real world at all.[35] However, the natural world offers experiences beyond anything a computer can generate. The computer can provide information about things people experience in nature, expanding on the initial stimulus. But all the scientific details are not necessary to enjoy sharing the experience of noticing, observing, and wondering about nature.

In all seasons, and all times of the day, awakening the senses to feeling the surrounding messages of nature stimulates imagination and establishes a sense of well-being and satisfaction. Noticing a familiar scene as it changes from season to season sharpens awareness and creates a connection to a specific place and time. Whether viewing a grand landscape or a small wonder of nature, seeing with an observant eye opens the world to a tapestry of wonders to enrich each day. Connecting to nature opens the mind to the beauty, majesty, power, and voice of the earth.

RECONNECTING TO NATURE: PITTSBURGH PATHWAYS FORWARD

Although we are unlikely ever to return to a time when we were "natural," our self-interest is gradually forcing us to develop a perspective of stewardship rather than dominion if we wish to survive. Today's growing environmental ethic exists because our impact on the globe is finally fouling our own nest.

MARK L. WINSTON

In the industrial period of Pittsburgh, clean air, water, and land were casualties of the economic priorities of industry. Throughout the history of Pittsburgh from the founding days in the 1750s and 1760s through the time of World War II, smoke symbolized the economic prosperity and vitality of heavy industries. Attempts in 1911 to abate the smoky air, notably headed by the Mellon family—bankers with interests in aluminum and oil rather than iron and steel—collapsed in the face of the press for advancing heavy industry.[36] The polluted air from steel and coke works became synonymous with economic prosperity, and the presumption of environmental degradation as a necessary sacrifice for economic prosperity continues to linger in the public mind. Even to this day people equate cleaning up the environment with losing good jobs.

The battle to mitigate smoke in the air was taken up in the early 1890s by elite reformers as a distraction from the problems of inadequate earnings, long hours, unsafe working conditions, and lack of unionization that concerned much of the populace.[37] The effort saw little success until after World War II through the 1960s. Then, the shift from bituminous coal to natural gas and cleaner combustion technologies for households, commercial firms, railroads and some manufacturing industries, combined with a municipal smoke control policy, eliminated the dense smoke from the atmosphere.[38]

The crash of the steel industry in the late 1980s was often blamed on the environmental protection laws that came into force in the 1970s. But in fact, the departure of heavy industry from Pittsburgh was more strongly influenced by market conditions and deliberate industrial decisions to move to areas where production and labor costs were much lower. This departure left former mill communities blighted with decades of disinvestment and decay. Providing safe environments and beautiful natural spaces in areas abandoned by the departure of industry still requires attention. Some industries that have become part of the current economy, such as coal-fired power plants, still impose disproportionately high environmental burdens on the communities close to them, and the battle for improved healthy air continues in Pittsburgh. However, many of the growing industries of modern times depend on cleaner high-tech businesses and knowledge-based industries and services. Today's workers in these industries expect to have living conditions that add to the quality of life, and that includes life outdoors. In Pittsburgh, more communities are organizing around improving the neighborhood as a way to begin creating renewed health for both people and the economy.

Pittsburgh has a renewed interest in restoring green spaces and defending small neighborhood parks from development. From small neighborhood green spaces to large city-owned areas, parks have been a part of Pittsburgh's heritage for a century. The four major parks that came to the city as gifts from the titans of industry in the early twentieth century now form the centerpieces of a new sense of value for natural spaces: Frick Park, Phipps Conservatory, Schenley Park, and Highland Park Zoo. Since 1996, other parks, originally established for water reservoirs or as city parks, are receiving renewed attention under the care of the Pittsburgh Parks Conservancy. New initiatives for green infrastructure to control stormwater runoff have had a particularly significant effect in adding green spaces such as rain gardens and bioswales within paved areas, and green roofs on city buildings. The natural elements of the environment are becoming more evident in urban gardens and in walkable paths and streets.

Preserving parkland has only recently emerged as a priority for Pittsburgh—market value has been the chief determinant of land use. But the terrain itself offers inherent resistance to the press of human development. Steep hillsides covered with trees separate densely developed corridors. Land depopulated by the downturn of heavy manufacturing at the end of the 1980s now offers spaces for reclamation and regeneration. Many of these areas have been reclaimed by natural succession of plants over rubble, but in some places, thoughtful and deliberate planning opens new horizons for people as well as living systems.

Pittsburgh today represents a totally different cityscape from the smoky, gritty, choking environment of the 1950s. Pittsburgh as a community is making a deliberate choice to include a safe healthy environment as a civic priority. Even as the press for new development goes forward, Pittsburghers are paying attention to creating or preserving natural spaces as a necessary part of urban life. These kinds of actions embody an ethic of care for the natural environment. Four case studies illustrate how placing priority on a healthy environment can affect choices in neighborhoods, schools, and communities. Examples from the Allegheny Riverfront Park, Magee-Womens Hospital, Green Building Alliance Healthy Schools Program, and the National Aviary Neighborhood Nestwatch program offer examples of an environmental ethic in action.

HEALING GARDENS AT MAGEE-WOMENS HOSPITAL

Sometimes a bit of nature in an urban space can have an impact beyond its small scale. Magee-Womens Hospital of the University of Pittsburgh Medical Center (UPMC) has been providing care to women and infants since 1911. The old blackened brick hospital next door to the nursing school dormitory has witnessed the arrival into the world of thousands of children in the greater Pittsburgh area: more than five hundred thousand babies have been born there, and outpatient visits have risen to two hundred thousand annually.[39] In addition to providing care to patients, Magee physicians and researchers contribute to the advance of science and innovation with a strong focus on research and education.

Judith Focareta, RN, Med., the hospital's first Coordinator of Environmental Health Initiatives, has set up a "Green Team" of physicians, nurses, community representatives, and environmental experts in a systemwide effort to study the effect of toxins in the environment on health. In addition to making significant procedural and systematic changes to reduce or eliminate toxins in hospital operations, education and outreach to patients was a significant part of the project.[40] The Green Team decided to use organic and healing gardens in three indoor courtyards at Magee as part of their teaching tool, Focareta explained:

> Two of the gardens furnish organic vegetables and herbs, which are utilized by our dietary staff in healthy meals for both staff and patients. The third garden is a sensory garden, which provides a relaxing space for staff, patients, and visitors. We recently offered tours to dietary staff of other UPMC facilities who are interested in following our lead. The use of pesticides is linked to cancer, endocrine disruption, and developmental delays. We advise our families to avoid pesticide use. It was a logical step to model that behavior. The gardens also serve as venues for both staff and patient education.[41]

Initially, nursing students from Carlow College volunteered to help with harvest and garden maintenance, but beginning in 2011, Magee has retained a full-time gardener on staff to coordinate the work.

A wide diversity of plants and wildlife grace these areas, including honeybees and songbirds. When I visited a friend at Magee in January 2014, several birds' nests from the prior summer were visible in the bare branches of the courtyard garden. The three Magee gardens, though occupying interior walled areas within the hospital, are open to the sky. Within the ten-block area of the mostly concrete, glass, and paved surfaces of the medical complex, they offer a welcome space of living earth. The gardens are positioned so they can be seen from patient rooms, especially in the cancer unit and the Woman Care Birth Center.

Benefits of having gardens within the hospital include stress reduction and a general sense of well-being. Focareta noted, "Numerous studies show that access to outdoor spaces reduces both length of hospital stay and the amount of pain medication required."[42] Access to the gardens can also have a restorative effect on patients' family members, as reported by the *Pittsburgh Post-Gazette*: "Nine-year-old Isabella Moosa of Export waited at Magee-Womens Hospital as her mother underwent surgery. Her elbows on a balcony railing, she overlooked a garden pond filled with large, brightly colored koi and big green turtles swimming in circles waiting for food to be dropped into the calm water. 'It's gorgeous,' she said. 'I love the way everything is full of life, especially the pond.'"[43] Moosa went on to add, "This is a place where people come to get help, and sometimes they are very sad. They need something to bring up their life. While they're here, they need some inspiration."[44]

Leslie Davis, CEO, said: "Magee is committed to providing superior healthcare to the families of western Pennsylvania and is a leader in advancing environmentally sustainable practices in healthcare, linking all aspects of our mission to greening initiatives and environmental health, safety, and stewardship."[45] Focareta explained that the Magee commitment to connecting a healthy environment to healthy people was inspired by the Women's Health and Environment conference of 2007 convened by Teresa Heinz.[46] Over two thousand women from all walks of life attended the daylong event funded by the Heinz Endowments and sponsored by Magee. As Focareta and her team began their green project, they had the freedom to undertake some unconventional activities not usually found in a hospital.

Throughout the growing season, the Magee Café menu includes fruits and vegetables harvested from the gardens. Scenes from the gardens and descriptions of the healthy attributes of the organic produce grace the Café décor, such as the poster proclaiming: *No pesticides served here!* Patient and visitor education includes showing people how organic food can reduce toxins in the environment and provides a place where the values of clean air and a natural environment can drive home the lesson that a healthy environment promotes healthy people.

ALLEGHENY RIVERFRONT PARK: BUILDING WALKABLE COMMUNITIES

Communities where walking and bike riding are safe and encouraged offer many more opportunities for people to experience the natural environment. Communities can plant city trees and create traffic-free zones to enhance this access and relationship to nature. Tree-shaded sidewalks or bike paths provide habitat and shelter for a larger variety of birds and wildlife than paved roads. This is especially effective if roadside plantings connect with green spaces such as parks or rain gardens. Moving at walking or bike pace offers a human-scale association with the natural world rather than the technology shaped experience of an enclosed vehicle. While speed, isolation from the elements, and privacy have become expected conveniences of modern travel, vehicles impede developing an integrated sense of community. Neighbors cannot talk to each other as they pass by in cars, but will often stop to converse if they are walking. Designing living spaces to include walkability and diversity of residential, commercial, and recreational spaces can increase the sense of community in a natural habitat for humans as well as increase the diversity of species living even in urban spaces.

A pedestrian park along the Allegheny River was first envisioned by Fredrick Law Olmstead as early as 1911.[47] Once relegated to an industrial transportation corridor for barges carrying coal, iron ore, and heavy cargo and the recipient of the effluent from factories, the Allegheny River was isolated from people on the shores. Instead of grasses and wetland shores, the edges were hard-cased in cement walls, docks, and piers with an oily slick on the surface that foamed and pooled in the eddies. People walking on the edge of the rivers in the 1970s risked being covered with black, greasy slime. This once-sterile scene has been transformed to a place accessible to people and one well suited to biking, jogging, dog walking, strolling with friends, or sitting to enjoy lunch and watch the ducks and birds. This linear park, with shaded walkways on the upper level and a lower level close enough to allow people to touch the water, streams with visitors and city residents at all seasons.

The Allegheny Riverfront Park runs along the south bank of the Allegheny River in downtown Pittsburgh.

ALLEGHENY CULTURAL TRUST

The concept of a riverfront park in Pittsburgh proposed the creation of two narrow strips of public land—one following the south bank of the Allegheny River and the other running along the north bank of the Monongahela River.[48] The proposal lay dormant until the early 1990s, when the Pittsburgh Cultural Trust's District Plan called for the creation of a riverfront park to border the northern boundary of the Cultural District. At that time, the Trust's Public Arts Advisory Committee commissioned a first-time collaboration between artist Ann Hamilton and landscape architect Michael Van Valkenburgh to create the Allegheny Riverfront Park.[49]

Cultivating living spaces within urban areas can enhance the sense of people being connected to the natural world. In addition to green islands as walkways and bikeways, Pittsburgh is increasing use of green roofs for energy conservation, water management, and to enhance natural habitat.[50] Green roof spaces appear on many business buildings in the city. Every flat roof provides a surface for some degree of green space to assist with energy conservation and water runoff management. Such spaces attract birds, insects, and other living forms as well. Visitors to the David Lawrence Convention Center in Pittsburgh can stroll or sit in a native plant garden that forms the roof of the lower level.[51] People who enjoy the river's recreational opportunities are better stewards of its health. Through the riverfront parks, the three rivers that define Pittsburgh also offer pedestrian-accessible connections between the life of the water as a habitat and the life of the city as a business and cultural center.

GREEN AND HEALTHY SCHOOLS ACADEMY: CREATING HEALTHY PLACES TO LIVE, LEARN, AND PLAY

Providing a stimulating and healthy environment for children in schools was identified as a strategic priority for the Pittsburgh Green Building Alliance in 2011. With a grant from the Heinz Endowments, the Green Building Alliance operates the Green and Healthy Schools Academy as an ongoing program to incorporate the principles of safe, healthy, and fun learning environments into schools.[52] The program brings the outdoors into classrooms and takes children outside in safe, accessible, natural environments. The Green and Healthy Schools Academy aligns with the US Department of Education Green Ribbon Schools award program to incentivize and recognize public and private K–12 institutions that work toward three pillars of a green and healthy school:

1. Net-zero environmental impact and reduced operating costs
2. Healthy environments and lifestyles for all students, teachers, and staff
3. Integrated environmental and sustainability education and civic engagement opportunities, which contribute to strong college and career pathways and twenty-first-century citizenship[53]

Jenna Cramer, program director for the Green and Healthy Schools Academy, explained the vision for the program:

> Children learn best in a space that is healthy and safe, that inspires creativity and confidence, and provides a culture for connecting to the natural world. We know that schools are complex institutions reflecting all the problems of the communities they serve. We hope the Green and Healthy Schools Academy can bring conditions where every child can attend an inspiring, engaging, healthy and high-performing school. The children are our future, they deserve the best we can give them.[54]

There are seventeen institutions in the Green and Healthy Schools Academy, ranging from very large school districts such as the City of Pittsburgh Public Schools to small private schools such as the Waldorf School. One of the partners in the program is Winchester Thurston School, a private pre-K through grade-twelve school with an urban campus in the City of Pittsburgh and a campus on eight acres of woods and meadows in the northern suburb of Fox Chapel.

Teresa DeFlitch, director of learning and innovation at Winchester Thurston, served as part of the leadership team for the Green and Healthy Schools Academy. The Winchester Thurston mission of engaged, experience-based learning uniquely incorporates a credo of ethics "that develops the mind, motivates the passion to achieve, and cultivates the character to serve. In addition, a rare ethos—over a century old—comes to life each day in our classrooms, as students learn to think also of the comfort and the rights of others. This mandate from our visionary founder is at the core of our intentional ethical development, our academic approaches, and our student life programs."[55]

One of the most interesting developments has been the City Campus Outdoor Classroom. Newly renovated in 2013, the Outdoor Classroom features a green-roof playhouse and a variety of gardens and planting spaces, and extends scientific learning to the outdoors. Here, students grow vegetables and learn about cultivating plants, water management, and healthy nutrition.

The connection with the natural environment infuses all aspects of academic life. In the summer of 2012, when the Outdoor Classroom concept was coming to fruition, DeFlitch explained that the students have environmental sustainability woven through the curriculum throughout their experience from kindergarten to high school with age appropriate experiences. On the eight acres of the Northern Campus, DeFlitch shared the vision of having specific outdoor spaces for learning experiences: "We encourage learning by doing, and it is important for children to experience their place in the natural world as they are shaping their own values. Beginning even in the pre-K levels, students find ways to understand systems, and how they are connected to the wider world."[56]

The experiential learning system has engaged students with the local water and sewer authority, ALCOSAN, to make an instructional video for use in outreach education. They have learned about air quality and health by working with the Group Against Smog and Pollution. DeFlitch notes, "The city really is our campus, and students engage at many levels. They graduate with a better understanding not only of their place in the natural world but with a commitment to take leadership in their communities."[57]

NATIONAL AVIARY NEIGHBORHOOD NESTWATCH PROGRAM

Bird watching is the second-most popular hobby worldwide, after gardening. It is available without much fuss. Just noticing and identifying common birds is fun for children. They can observe the competition among the different individuals, and the antics of the squirrels and chipmunks around a birdfeeder. If there is an opportunity to watch a nesting cycle, children relate to the parent birds' nurture and fledging of chicks. Birding plays an important role also in tracking the changes in the earth's climate and in the effects of human development on habitat.[58] Learning about bird migration gives people the perspective of these world travelers on the interconnectedness of all living things. Avian visitors to backyards and city trees can only find shelter, food, and resting places if people take care to preserve open spaces and hospitable habitats for them. Understanding the journeys of bird visitors and residents can open a world of delight to children, and can form a lifelong interest. Bird watching also bridges generations as people with long Bird Life Lists share their knowledge with others.

The National Aviary, located in Pittsburgh, participates in a nationwide study of migratory bird patterns. Dr. Peter Marra at the Smithsonian Institute in Washington, DC, developed the Nestwatch program in 2000 to help scientists better understand the ways birds are adapting to urban environments.[59] This program uses the participant-observer capacity of citizens resident in urban and suburban communities to track migrating birds. In 2013, the National Aviary began a citizen observer Neighborhood Nestwatch program engaging two hundred locations in urban, suburban, and rural areas to track eight common bird species through a nesting season. A team of ornithologists visits a backyard to safely capture, record vital statistics, and uniquely band the birds with colored plastic markers and an aluminum numbered band.[60]

By its design and the nature of the activity, Neighborhood Nestwatch creates a bond between people and the migrating birds. Families are encouraged to participate, especially in releasing the birds after they are recorded and banded. Participating in a Neighborhood Nestwatch program can encourage children, from an early age, to hear and distinguish the birds of their own neighborhood. Providing a feeder can draw many birds close enough to view. Learning birdsongs and seeing the different species that appear at different times of year can introduce children to a wide world. Neighborhood Nestwatch came to the home of a family in the Blawnox neighborhood, overlooking the Allegheny River in May 2014. The grandmother, parents, and five children ranging in age from five months to eight years participated in the process of spotting the birds and helping to release them after they were tallied and banded. When it was her turn to release a bird, four-year-old Stella was so excited she could hardly talk until after the American song sparrow took flight from her little hands. Then she was jumping up and down. "The birdie was so soft to hold," she said.

Children, neighbors, and friends can make a day of helping the process and learning to identify and spot the target species of the program: American robin, gray catbird, northern mockingbird, Carolina wren, house wren, black-capped/Carolina chickadee, northern cardinal, and song sparrow. Each site is named, and given a tracking sheet on which to record and report sightings of birds nesting, the date of each stage from egg laying to fledging, and the fate of the nest.

As the years progress, banded birds often return to the same nesting location, sometimes with the same mate, and sometimes with new pairings. Children and adults alike develop a very personal interest in the banded birds and can begin to notice the particular behavior of individual birds. For those that migrate, like the American robin, the return of a banded one each spring becomes an occasion for celebration. The first bird banded in the Nestwatch program in Pittsburgh is an American robin, named "The Master of the Garden" with his Pittsburgh black and gold bracelets. He has returned for four years to take up residence in a little patch of Forest Hills. The forces have aligned to allow this little traveler safe passage away and back home. People who never noticed birds before develop an avid interest in the care of those they learn to watch over. They are less likely to use harmful chemicals in their property, and will take efforts to make a diversity of habitat to attract birds. The process enriches lives and enhances the connections between people and the living earth.

CONCLUSION

People preserve and nurture what they love. Infusing a sense of connection and love for nature builds a sound foundation for the decisions individuals will face throughout life regarding how they relate to the natural world. As a society, investing in the preservation of the living earth and its healthy ecosystems can build resilience in the face of change. People, schools, communities, and businesses thrive when the natural environment is healthy. Perhaps in the twenty-first century a flourishing natural environment, lush and green with life, fresh air, and cool clean rivers, will displace the smoke of the Industrial Revolution as a marker of Pittsburgh's prosperity.

The Esther Barazzone Center at Chatham University's Eden Hall Campus.
CHATHAM UNIVERSITY

PART II

CHOOSING SUSTAINABLE PATHWAYS

Economic revolutions don't just emerge from the ether. The laying down of a new communications and energy infrastructure has always been a joint effort between government and industry.

Our economic priorities need to shift from productivity to generativity, and from a purely utilitarian pursuit of nature, to stewarding the relationships that maintain the biosphere.

JEREMY RIFKIN, *THE THIRD INDUSTRIAL REVOLUTION: HOW LATERAL POWER IS TRANSFORMING ENERGY, THE ECONOMY, AND THE WORLD* (NEW YORK: PALGRAVE MACMILLAN, 2013), 129, 225

Windstax is an innovative approach to wind power developed by owner and founder Ron Gdovic in 2012.

WINDSTAX

4

TRANSFORMING THE ENERGY SYSTEM

Unlike fossil fuels, sunlight is a flow and not a stock. Once a gallon of oil is burned, it is gone forever, but the sun will cast its rays earthward a billion years from now, whether sunshine is harnessed today for human needs or not.

DENIS HAYES

AS THE INTERNATIONAL preparations for the Twenty-First Conference of Parties (COP21) began to include advance commitments for action on climate change, President Obama, through the existing powers of the Environmental Protection Agency (EPA), proposed a Clean Power Plan (CPP) with specific carbon dioxide emission reduction goals. In the summer of 2014, the EPA conducted a series of hearings on the proposed plan designed to reduce emissions from electricity generation stations, the largest contributors to climate change. This regulatory initiative was intended to be the core of the US response to global climate change. The plan's stated national goal for 2030 was to "cut carbon emissions from the power sector by 30 percent from 2005 levels, while starting to make progress toward meaningful reductions in 2020."[1]

The CPP required each state to develop its own plan within the national framework.

Among the hearings the EPA conducted across the country, none more sharply defined the conflicting arguments than the one held in Pittsburgh. The voices of the coal and fossil industry advocates with over 3,500 members from the United Mine Workers, International Brotherhood of Electrical Workers, United Steel Workers, and other unions, objected to the "war on coal" and clamored for an end to regulatory limits on the industry. Their numbers filled the streets in a river of outrage at what they saw as a challenge to their way of life and the long, proud tradition of coal mining and use that once defined the industrial prowess of Pittsburgh and America. Clustered on a corner plaza a block away from the Federal Building, a few hundred environmental activists held signs and chanted for action on climate change. In the lead was a brigade of mothers with babies in strollers along with volunteer members of the Sierra Club, Marcellus Outreach Butler, the Group Against Smog and Pollution, 350 Pittsburgh, Penn Future, Penn Environment, Women for a Healthy Environment, and many others holding handmade signs and an assortment of T-shirts and slogans pleading for clean air and solar power. As the two groups shouted at each other, labor historian Charles McCollester stood at the intersection with his own handpainted sign that stated: *When blue collar workers fight green earth health we are all doomed!* The passion of both sides came down to fears for the future; on the one hand, coal miners worried about providing for their families and their own pensions and health benefits in the absence of the industry that supports their jobs, and, on the other hand, those pleading for healthier conditions for children of the future and the stability and health of the earth. The two groups were not speaking the same language, nor were the demonstrators crossing the divide to examine the reality each side faced.

The ethical issues inherent in climate change rest most heavily in the energy system. Energy runs modern life, and the fossil-fueled energy system accounts for most of the greenhouse gases causing climate change. Therefore, as more people accept the urgent need for climate action, renewable and sustainable energy systems are receiving more attention. Solar power, wind power, and advanced energy storage systems have become more cost-efficient and more widely adopted. Yet the shift in the energy base of the economy from fossil fuels—coal, petroleum, and natural gas—to renewable and sustainable systems moves unevenly as different areas make adaptations or adjust to new opportunities.

In the absence of congressionally authorized policy changes to address not only the reduction in use of fossil fuels but also the social, economic, and community adaptations necessary for a smooth and just transition, regulatory actions to reduce fossil fuel combustion are characterized as "a war on coal." The transition to a renewable and sustainable energy system is proceeding not by measured and deliberative actions with a comprehensive policy but as a series of battles pitting fossil fuel interests against the proponents of environmental preservation. The resulting strife creates a paralysis of policy and a pitched battle between unequal combatants wherein the long-term public interest and the preservation of nonhuman life fall outside the lines of the debate. With the election of Donald Trump as president in November 2016, the removal of environmental regulations and explicit promotion for coal, oil, and gas development press federal policy further in favor of the fossil fuel industry, making the advance of more sustainable policies even more difficult.

There has been no broad consensus of public thought on how to address the looming threat of climate change or the ethical problems it poses. The elements of the transition have been analyzed and solutions proposed in think tanks and laboratories, but government institutions have not placed a priority on the outreach to and education of the broader public. Instead, the fate of the living planet and the next generation yet to come hangs on the forces that control political power.

Power and fuel sources are, for the most part, out of sight and mind unless there is a sharp increase in price or an outage from a passing storm. The uncoupling of fossil fuel consumption from economic growth has begun; numerous initiatives toward sustainable energy systems are proceeding with increasing momentum. The technologies necessary for a new energy system transition already exist and grow more robust as they begin to penetrate the market. However, the institutional framework of the energy system is deeply embedded with the fossil system. The necessary transformation of how people everywhere power the functions of civil life requires a broad consensus of public opinion and a wide empowerment for a change in the direction of energy policy.

THE EVOLUTION OF AMERICAN ENERGY POLICY

In the seventeenth century, foundries and early iron works were fueled by wood and charcoal, with hydropower and windmills producing mechanical shaft rotation power to run sawmills, grain mills, and the carding and spinning mills of the textile industry. By the middle of the 1800s, however, new inventions propelled this industrial profile into high gear, and the human enterprise began to change and shape the land in much more permanent and drastic ways. In colonial times, deforestation for timber and land clearing for agriculture and road building characterized the human presence on the land, but the advent of coal mining, burning, and waste deposition that propelled the Industrial Revolution in the 1900s changed the landscape on a much grander scale. "The black rock dug so easily from exposed seams yielded a fuel far superior to the wood and charcoal on which the [early] Americans had to rely while living on the Atlantic Coast. Throughout the nineteenth century, coal fired the iron furnaces of Pittsburgh, heated the factories of New York, and drove the locomotives carrying settlers west across the continent."[2]

This period is important to understand because persistent national attitudes toward resource use were laid down with the westward expansion of the nation. The resources of the continent were defined and valued in relation to their usefulness to the economic enterprise. Energy policies, to the extent they existed at that time, addressed access to secure lands for westward expansion and to secure development. Some of the policies adopted in the 1800s remain in effect today, such as below-market price leasing of public lands to encourage mining and exploration. The spiritual and aesthetic value of the land, more deeply embedded in the cultures of the indigenous peoples of America, were expounded in Euro-American cultures in the writings of David Thoreau, George Perkins Marsh, John Burroughs, and John Muir.[3] However, the first national conservation policies were not established until Theodore Roosevelt's administration.[4]

Energy policy in the industrial expansion period of the late nineteenth century into the twentieth century evolved to form a complex mix of state and federal regulatory policies mostly designed to assure the flow of energy to keep the economy rolling. Cheap fuels drove the economy, and were highly subsidized. Tax subsidies for oil and gas exploration and for investment in coal and production facilities were laid down in law to stimulate business, to support production expansion in wartime, and to protect domestic industries from foreign competition. The infrastructure of roads, pipelines, and energy production and distribution facilities spread the advances of the Industrial Revolution nationwide.

By the 1900s the electricity business, started with Thomas Edison in 1882, had expanded to a plethora of small companies, some private and some municipal. World War I increased electricity use for national defense purposes and the federal government received authority over large hydropower development through the Federal Water Power Act of 1920. As the industry continued to grow, utility companies extended their reach into multiple geographic areas by establishing holding companies that bought up the smaller utility companies across the country. Eventually, technology advanced for transporting large quantities of electricity greater distances, and early analog computers allowed remote switching and controls. Central power stations serving remote customers were established.[5] By 1935 the need for some sort of federal regulation of the energy industry became apparent, and a system of franchises, regulated by the states as to price and performance, came into being through the Federal Power Act of 1935 and the Public Utility Holding Company Act.[6] Wholesale utility transactions and transactions between states were regulated at the federal level by the Federal Power Commission, later the Federal Energy Regulatory Commission. State utility commissions governed internal utility company transactions and tariff rulings within states.

After World War II, consumer spending was encouraged to help the recovery, and conspicuous consumption became synonymous with prosperity. Declining block electricity pricing—in which the more electricity customers used, the less they paid per unit—encouraged electricity use for labor-saving appliances. Banks gave away electric toasters and small appliances as premiums. People drove everywhere for work and pleasure as the automobile spread across the country and settlement fled the industrial cities to settle in sprawling suburbs.

It all came to an abrupt halt in 1973. The early 1970s were tumultuous years with social upheaval rampant around issues of women's rights, racial discrimination and voting rights, environmental protection, and opposition to the unpopular Vietnam War. On top of all of this turmoil came the Arab Oil Embargo from October 1973 to March 1974, in protest of American support for Israel in the Yom Kippur War.[7] Oil went from a commodity that was taken for granted to the focus of a crisis that touched everybody. Most people had never really thought about where energy came from, but this sudden experience awoke national sensibilities. Energy policy initiatives changed to include national security considerations and environmental impact.

The initial responses to the energy crisis came in President Nixon's call for American energy independence, with the eventual enactment of the National Energy Act of 1978 that laid out a blueprint for reducing American dependency on imported oil, among other things. In addressing the nation on the introduction of the legislation on this energy crisis, President Nixon said:

> As far as energy is concerned, this means we will hold our fate and our future in our hands alone. As we look to the future, we can do so, confident that the energy crisis will be resolved not only for our time but for all time. We will once again have plentiful supplies of energy, which helped to build the greatest industrial nation and one of the highest standards of living in the world. The capacity for self-sufficiency in energy is a great goal. It is also an essential goal, and we are going to achieve it.[8]

This initiative passed in Congress with bipartisan support and was carried forward by three presidents—Nixon, Ford, and Carter—placing America on a policy path to a more diverse, less wasteful, and more secure energy system.

Fear of energy resource scarcity in the face of growing global demand propelled the energy independence initiative. President Carter tried to shift the focus of US energy policy from not only increasing domestic supplies but cutting waste and controlling demand. In a 1977 Oval Office address, delivered in a cardigan sweater before a fireplace, he spoke of the need to find alternatives to oil: "Our decision about energy will test the character of the American people and the ability of the president and the Congress to govern. This difficult effort will be the 'moral equivalent of war'—except that we will be uniting our efforts to build and not to destroy."[9] From 1978 to 1984, the gross domestic product rose by 11 percent while energy use per capita fell by 10 percent, illustrating an increase in productivity and energy efficiency across the economy.[10] America's economy was moving in a more energy-efficient direction.

The election of Ronald Reagan in 1981 ushered in a change in the winds of public opinion, and public policy actions moved away from government controls in the public interest toward market-based actions. President Reagan explained this philosophy:

> The National Energy Policy Plan that I am sending you, as required by Section 801 of the Department of Energy Organization Act (Public Law 95–91), represents a break from the format and philosophy of the two National Energy Plans that preceded it. Our national energy plan should not be a rigid set of production and conservation goals dictated by Government. Our primary objective is simply for our citizens to have *enough* energy, and it is up to them to decide how much energy that is, and in what form and manner it will reach them. When the free market is permitted to work the way it should, millions of individual choices and judgments will produce the proper balance of supply and demand our economy needs.[11]

The market signals around energy included significant permanent subsidies designed to stimulate domestic production of oil, natural gas, and coal, but the Tax Reform Act of 1986 removed the tax deduction for energy efficiency improvements for homeowners and small businesses and cut subsidies and supports for renewable energy, though existing subsidies for oil, gas, and coal development remained. The budding domestic solar industry crashed even as President Reagan symbolically removed from the White House roof the solar system installed under President Carter. Funding for public information programs about energy use, conservation, and the consequences of pollution were also significantly reduced or removed from federal budgets. Thus, at the same time that market forces became the primary impetus in energy use decisions, consumer information concerning such market energy choices came less frequently from government programs and more frequently from volunteer nonprofit organizations or marketing messages from energy industries.

The shift to a market-driven energy policy accelerated American use of fossil fuels with increasingly destructive and invasive extraction technologies. The proportion of US dependency on imported oil at the time of the Arab Oil Embargo in 1973 was 37.3 percent; by 2010, imports of oil reached 70 percent, even with vastly increased domestic production.[12] Coal fueled approximately 50 percent of the electricity in the United States, much of it mined from thin seams by mountaintop removal.[13] In this period US carbon dioxide emissions began its rapid upward escalation.

This era of market-driven policy initiatives also marked the beginning of significant political polarization based on ideologies. The power of scientists to influence public policy began to diminish. Movement to "get government off the backs of business" inspired the deregulation of airlines, telecommunications, natural gas distribution, and finally electricity.[14] The policy directives in energy were increasingly delegated to the states for final implementation, especially after the Energy Policy Act of 1992 opened the way to energy industry restructuring. Under Presidents George H. W. Bush and Bill Clinton, the formerly integrated electric utility industry, where regulated monopoly companies operated generation, transmission, and distribution of power within a given franchise area, was disaggregated into separate entities, with varying regulatory jurisdictions.

Under electricity deregulation, customers were given the ability to choose among energy service providers. The states were given authority to design the deregulated energy markets and set their own systems in place to respond to the market competition provisions of the law. Consumers were faced with a bewildering array of information, and many decided to remain with the default provider rather than try to switch companies.

California led an initial flurry of interest wherein states adopted consumer choice options in a system that uncoupled electricity generation and transmission from the local distribution and billing functions. The partially deregulated market established in California left room for energy traders to game the system, creating an artificial energy scarcity to drive up their profits. In 2000 and 2001, the failure of the California energy markets with rolling blackouts and scandals in the energy-trading arena created chaos.[15] California's experience brought a sense of skepticism to many states that had not completed action on deregulation.

After a decade of state energy regulatory restructuring proceedings, in 2012 and continuing to the present, there are sixteen states with fully competitive markets, twenty-seven with traditional regulated electricity industries at the state level, and seven where initial deregulation initiatives were suspended or curtailed.[16] For detailed conditions in each state, see the Database of State Incentives for Renewables and Efficiency.[17] The struggle to transform the energy system in response to climate change will play out in this many-layered, complex regulatory framework.

More than half of the greenhouse gases from human activities that are causing climate change are created by transportation and electricity generation.[18] Yet these two sectors, driven primarily by technologies developed in the Victorian age, are also the most wasteful. Electricity-generation technology, based on the Rankine steam engine invented in 1866, converts only 33 to 36 percent of the energy from fuel to electricity; the rest is lost as waste heat.[19] Over 90 percent of the electricity used today relies on this system. Meanwhile, modern transportation systems still operate primarily on the internal combustion engine, invented in 1865, which converts only 12 percent of the energy from the fuel it uses into forward motion; the bulk is lost as friction and heat. As a result, the energy system that supports the current economy wastes quantitatively more energy than is actually used to perform work. In 2015 the United States used 97.5 quadrillion British thermal units (quads) of energy from all sources to do all of the work of the economy.[20] Of that amount, only 39.5 percent (38.4 quads) went to useful work in residential and commercial buildings, operating machinery, factories, and transportation. The majority, 60.4 percent (59.1 quads), was lost as waste heat and friction; and 6 percent of fossil resources, coal, oil, and natural gas, were used directly as raw material for plastics, pharmaceuticals, chemical fertilizers, and synthetic products.[21]

Beyond the inefficiency of conversion from fuel to electricity, these energy processes also place a tremendous burden on the water supply. According to the US Geologic Service, thermoelectric cooling of power plants uses 45 percent of available ground and surface water in the United States.[22] Water availability and distribution, especially for surface water, is expected to be adversely affected by climate change. Many areas of the country already experience droughts and major shifts in precipitation patterns attributed to climate change. Priority uses for food production and sanitation will compete with this large demand by the electricity-generation sector.[23]

With the election of Donald Trump, US energy policy is shifting once again toward a fossil-fueled basis, with greater emphasis on domestic development, especially coal. Rick Perry, former governor of Texas, was appointed Secretary of Energy and stated in his confirmation hearing:

> I believe the climate is changing. I believe some of it is naturally occurring, but some of it is also caused by manmade activity. The question is how do we address it in a thoughtful way that doesn't compromise economic growth, the affordability of energy, or American jobs. . . . But we truly advocated an all of

the above strategy, reducing carbon emissions not just through development of cleaner fossil fuels, but through the development of renewable sources too. During my time as governor, Texas took the national lead in wind energy development, and now produces more wind power than all but five countries. When it comes to climate change, I am committed to making decisions based on sound science and that also take into account the economic impact.[24]

THE POSSIBILITIES OF RENEWABLE ENERGY

New ways of using energy capturing sun power can create less toxic, less damaging, and less destructive ways of doing work. A renewable resource is naturally regenerated over a short time scale and derived directly from the sun (such as thermal, photochemical, and photoelectric), indirectly from the sun (such as wind, hydropower, and photosynthetic energy stored in biomass), or from other natural movements and mechanisms of the environment (such as geothermal and tidal energy). These renewable energy sources provide a flow of energy falling on the earth as a condition of the planet's existence in the solar system.

The work currently done by burning fossil fuels can be performed through use of renewable energy systems that are more consistent with the natural ecosystems of the living earth. Renewable energy systems can be aligned with the location, quality, and amount of energy to perform specific tasks like space conditioning, lighting, appliances and machinery, and transportation. However, renewable energy continues to face skepticism engendered by unfamiliarity and misconceptions. "Is there enough renewable energy to replace cheap fossil fuels?" "What happens when the sun goes down or the wind stops?" "The high cost will kill jobs!" "Renewable energy uses rare materials that will be scarce and too expensive." These sentiments and concerns have good answers, and conditions improve every day. In the words of Leo Gerard, president of the United Steelworkers International Union, "The choice is not between good jobs and saving the environment. We must have both, or we will have neither."[25]

Renewable energy resources are abundant and ubiquitous. Renewable energy sources will replace an energy economy based on fear of scarcity to one grounded in the security of abundance. The largest energy source on earth is the sun, whether in fossil form of ancient trees and plants or as incident energy of sunlight, the potential energy from waterfalls, or biomass from growing plants. The sun produces energy from a continuous nuclear fusion reaction ninety-three million miles away from earth and delivers 8.2 million quads per year to the surface. The International Energy Agency estimates the global annual energy demand at eight hundred quads by 2050, even with modernization of emerging economies.[26] The absolute amount of renewable energy falling on the earth far exceeds our use or needs. It is already distributed around the globe, with a regular twenty-four-hour cycle. The winds blowing by the mechanics of the Earth's rotation are more variable, but also globally distributed, and the core earth-to-surface temperature gradient is constant. Figure 4.1 illustrates the availability of renewable energy flows far in excess of estimated demand now and for the future.

Renewable energy costs are becoming fully competitive with fossil fuels. The economy of fossil energy includes significant taxpayer subsidies. From the earliest trade supports in 1786, coal, oil, and natural gas have received taxpayer subsidies averaging $4 billion per year in the United States. On an international pretax basis, subsidies for petroleum products, electricity, natural gas, and coal reached $480 billion in 2011.[27] By comparison, taxpayer subsidies to renewable resources have averaged $400 million annually since 1994, with the levels and duration varying according to the whims of Congress. Action to

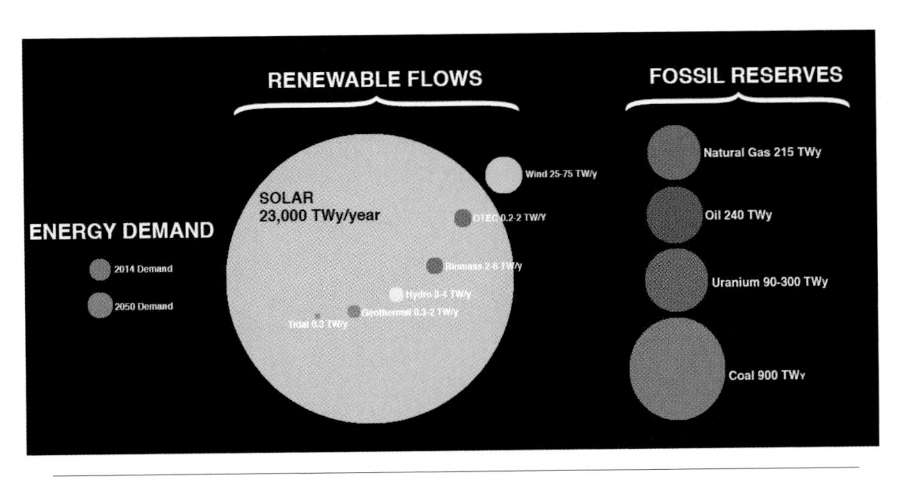

Comparison of fossil and renewable energy resources and demand to 2050.
BRITTANY SHEETS

retain such supports as production tax credits and investment tax credits for renewables face annual struggles against the fossil interest, for whose companies such supports have been fixtures in the budget for decades, regardless of the state of the company's need.

As President Obama pointed out in urging Congress to rescind oil and gas company subsidies, "Instead of taxpayer giveaways to an industry that's never been more profitable, we should be using that money to double down on investments in clean energy technologies that have never been more promising—investments in wind power and solar power and biofuels; investments in fuel-efficient cars and trucks, and energy-efficient homes and buildings. That's the future."[28] Obama included this proposal in every budget he submitted to Congress, but it has never received favorable action. Across the globe, rescinding price support subsidies for fossil fuels has proven extremely difficult because of fears that the cost of necessary energy commodities will fall disproportionately on those least able to pay, and because of the tremendous influence the fossil industries exert on the political process. Thus, the pricing structure created for fossil fuels has become a barrier to moving to a sustainable base.

Nevertheless, even in the absence of sustained and constant subsidies for renewable energy, the price per unit has fallen steadily over the last twenty years. According the National Renewable Energy Laboratory analysis in 2015,

> Renewable energy costs have declined dramatically over the past decades—some up to nearly 80% since 2009—with the most significant improvements in wind and solar photovoltaic. Cost reductions are likely to continue as markets expand and technologies improve. The levelized cost of electricity of solar photovoltaic energy dropped by 50% between 2010 and 2014 alone, making it increasingly competitive in both the residential and utility markets. Today, onshore wind levelized cost of energy falls either within the range or below the cost of new fossil fuel power generation.[29]

Renewable resources have minimal environmental damages. One other major influence on the current energy policy scene emerged from the National Energy Act of 2005, enacted under President George W. Bush and Vice-President Dick Cheney. This initiative again reduced focus on efficiency and renewable energy directions in favor of amplifying domestic oil and gas development, including offshore drilling, federal lands leasing, and a particular exemption for hydraulic fracturing for oil and gas. What came to be known as the "Halliburton Loophole" grants exemptions from seven major federal statutes intended to protect public health and the environment, including the Safe Drinking Water Act, Clean Air Act, and Toxic Release Inventory provisions.[30] However, ecosystem services used in the extraction, production, and use of fossil fuels are not counted in the price. The water polluted by coal mining or hydraulic fracturing for natural gas is not reflected in the price. The price of oil does not reflect the cost of cleaning up the thousands of gallons of spills each year.[31] None of these fuel prices reflect the impact of their combustion on the land, air, and water in terms of health or environmental degradation. None of the fossil fuel prices reflect the toll their combustion is taking on the stability of the climate, or the biodiversity of species.

Wind, solar, and hydroelectric energy technologies generate electricity with no combustion emissions. Wind and solar technologies also use no water to operate; thus they cause no pollution to water supplies or competition for other uses such as agriculture. While the production of renewable energy machinery has some environmental emissions, these are small compared to the mining and extraction of coal, oil, and natural gas, or the construction and operation of fossil-fueled power plants. The environmental impact of batteries associated with renewable energy systems has been noted as a source of environmental contamination. But experimentation in battery and storage technology is still in comparatively early stages. As applications of solar photovoltaic systems increase, the adaptive use creativity will advance, just as automobile designs have changed over time.

The work of the Rocky Mountain Institute over the last forty years has tested and proven that options for converting our society from a fossil-fueled base to a renewable and sustainable base are abundant and effective, even with current technology.[32] Noncombustion processes for generating power are available commercially and gaining in sophistication and effectiveness. With supportive policies, these minimally polluting technologies will evolve and become integrated into efficient systems. The barriers to pursuing a renewable and sustainable energy system drawn from the abundant flows of resources rather than from those bound as fossil deposits in the crust of the earth are not technological. They are political and ethical barriers. Renewable resources are not scarce; what is missing is the societal organization to tap the free, renewable energy flows in appropriate ways, and to link its capture and delivery system to energy end uses.

CHALLENGES TO A RENEWABLE ENERGY SYSTEM

Reaching a zero fossil-fuel energy system by 2050 will require changes in the way energy is procured, produced, and distributed. At the average rate of one percent per year increase in renewable resource use, the international goal of averting serious consequences for the living earth will not be reached.[33] If society is to reach the goals of moving away from fossil fuel dependence by 2050 and building sustainable, resilient communities, changes in the concept of utility services will be required. Many technologies are becoming available to drive a new utility business model that is compatible with system solutions. However, solutions based on integrating energy, water, and waste infrastructure with community needs face significant barriers. The markets and regulations governing utility services are embedded in national laws that favor existing fossil-fuel-based systems, making the pace of movement to non-fossil-combustion-based energy systems very slow.

How do we reach a renewable and sustainable energy base for our economy from a current status of 88 percent dependence on fossil fuels in a system established for two hundred years? In 2014 and 2015, the Institute for Green Science convened two roundtable workshops on utility and renewable resource interface. The first explored ways to accelerate rapid movement to an energy system that will meet sustainability goals.[34] This roundtable drew together twenty people from gas and electric utility companies, solar and wind manufacturers and distributors, conservation and efficiency service providers, and academic researchers. After a brief summary of the IPCC data, the workshop addressed the proposition: What actions would be necessary to accelerate the use of renewable energy systems as much as possible? The second roundtable convened the same group with the addition of experts on fuel cell technology and advanced battery storage systems. This session addressed the issue: How can a distributed renewable energy system operate with high reliability and twenty-four-hour service? Roundtable participants identified ways in which existing infrastructure can be used to support and shift to a system based on distributed and renewable resources, and ways in which that same infrastructure acts as an impediment. The combined results of these discussions revealed several areas for further action.

Pennsylvania is part of the Northeast Corridor, served by twenty natural gas delivery pipelines.[35] It is also a "customer choice" state, in which the customer can select a company to provide electricity generation. Choices include the traditional providers with a mix of coal, nuclear, and natural gas generation, as well as providers that sell renewable energy in proportions ranging from 20 to 100 percent, some generated exclusively in Pennsylvania. Most customers make their choice based on price.

Energy customers are buffeted by a broad array of choices, but they really enter the decision-making market at only a few critical times: at the point of purchase for a house, vehicle, or major appliance, or when they see a marked change in the price of energy. Most of the time, consumers do not want to think about energy, they just want everything to work on demand, all the time.

Because customers make their choices on the prices they see in the market at any given time, renewable energy choices currently suffer from the low price of natural gas from the Marcellus Shale boom. The advent of shale gas not only has shifted the electricity market from coal to natural gas but also has taken the focus of policymakers toward shoring up shale development options, reducing attention to renewable resource development. One roundtable participant noted: "We have large banks and foundations and universities with intellectual capital in Pittsburgh, but these assets are not specifically directed to cultivating renewables—they are distracted with Marcellus Shale and other things like carbon sequestration."

For a market-based energy system to work effectively, the rules and conditions need to be predictable and stable over time. Everyone at the roundtables agreed on the need for stability in incentives and market structure to allow for investments in renewable energy systems. With investment tax credits, production tax credits, and customer subsidies in the form of favorable financing options now highly variable and uncertain from year to year as state and federal policies change, businesses and customers both are groping for stability. The best way to accelerate the use of non-fossil-fuel energy resources is to stabilize the investment and production tax credits for renewable energy.

Participants in the roundtable explored ways to use the natural gas system as a bridge to non-fossil-fuel and noncombustion electricity processes. Technologies such as fuel cells and biogas production are readily available and are used in other parts of the world.[36] Fuel cells, as part of a customer's energy system, can become a bridge technology if fossil-fuel-sourced natural gas is replaced by biogas from anaerobic digestion of waste. Large-scale biogas generation faces the additional infrastructure hurdle of requiring integration with waste and wastewater handling systems. Some experience with biogas from anaerobic digestion of manure at dairy farms has been demonstrated in Pennsylvania where the gas drives a regular turbine generator to power the milk production equipment and to heat the barns.[37] Some communities are looking at their sewage treatment systems for anaerobic generation of methane to power the sewage treatment plant itself. Such biogas systems for generating methane can serve as backup for renewable energy sources with intermittent output.

To make this kind of integrated energy system work best, the separations between electric, gas, and water utility services would need to be addressed. While technology is converging in systems, the regulatory structure remains in isolated silos—gas, electricity, water, and sewer are regulated separately. Regulatory barriers come from the way the utilities are treated in the public utility system. The same customer may pay a gas utility, an electric utility, and a water and sewer utility for services. Tariffs for each of these services are set according to regulatory conditions established in the laws and regulations covering them in each state.

Blending solar/wind/biogas and rechargeable batteries as backup systems combined with high-performance buildings is a new challenge to the existing system of utility services. There is no tariff in place to address the price of biogas made from wastewater from a water/sewer utility. The infrastructure for connecting gas and electric services as integrated energy systems does not exist in the US regulatory system. At this time, for example, using gas-powered fuel cells to generate both electricity and space conditioning would make the gas utility a competitor to the electric utility, and the integration of the two has no smooth path. The existing process would involve contractual arrangements between the gas utility and the electricity utility.

As "intelligent buildings," or high-performance structures, in which utility services are integrated, become more prevalent, the separate control of different energy utilities becomes even more of an impediment to progress.[38] If a net-zero energy building has a gas-using power and space-conditioning unit that uses gas from a biogas source, the infrastructure to support such a system would have to be provided entirely by the customer's investment. This limits such highly efficient and effective systems to wealthy individuals, businesses, or institutions. There is no mechanism for different utility providers to invest in a net-zero energy/water/waste construction as part of the existing rate and tariff structure. If communities in redeveloping or putting up new developments strive to have a zero-net solution on a communitywide basis, they are currently obliged to be "off the grid" because there is no way for the utilities to become coinvestors in the necessary infrastructure with investments in the rate base. They can compete with each other to provide services, but there is no way for multiple utilities to invest jointly in innovative infrastructure within the current regulatory process.

Technology for customer-based energy such as rooftop solar photovoltaic energy and small-scale wind generators as well as advances in battery technology create customer options for self-generating some or all of their own energy needs. Electric vehicles with battery capacity allow customers to enter and exit the grid in ways that have not been modeled in the past. However, today's energy infrastructure is not designed to support intermittent renewable resources distributed among the customers and mixed in with the load.

The existing electric system generates power at large stations and distributes it to distally located loads of residential, commercial, or industrial customers. In Pittsburgh, the power supply infrastructure has excess capacity for current demand due to the historic design for steel mills and other heavy industrial uses. The predominant mode of electricity generation comes from coal plants, some in operation for over fifty years. Similarly, the natural gas distribution system has fourteen thousand miles of pipelines connecting eight thousand feed-in points from local gas producers to approximately seven hundred thousand customers. Most of the local gas suppliers are in small, single-source, shallow wells or farms, with a few from the Marcellus Shale formation.

Having electricity generation systems distributed among the load changes the utility function to one of aggregating the generation to meet the total load. For example, a solar hot water unit powered with photovoltaic cells on a customer's house would be programmed as a negative demand device, with a variable load over the twenty-four-hour day, and with seasonal variation in output. This customer demand element (hot water) would act more like a generator than an appliance. Because the cost of such technologies is continuing to fall, the limits on installation come from the regulatory interface with utilities, and from the financing of the upfront capital costs.

In an ideal situation, the utility tariffs for service are based on the costs of service to particular classes of customers—typically residential, commercial, and industrial, plus special contracts for unique customers. When the cost parameters are shifting from the customer's ability to both buy and sell electricity, traditional ways of computing the cost of service are no longer accurate. Changing the utility-planning model from central station sources to distributed generation among load centers entails a different business model for utility services.[39] Such models would need to be adjusted to include options for customer electricity generation, not only customer loads.

Providing basic utility services as a public obligation wherein all share in the costs of service within rules of fairness to both customers and to the shareholders of the investors in the utility is a central part of our social compact. As diverse producers and storage systems are added to the system, and incorporating the "smart meter" and "smart house" technology for controlling and monitoring load, the nature of the grid will change. The current system is becoming out of date for the rapidly expanding uses. The new technologies challenge the customary role of the public utility as the provider of last resort, with the obligation to serve all customers, even those who generate some of their own power. If large parts of the customer base drop off the grid to self-generate, what becomes of the legacy customers, and the existing system? These issues are at the forefront of industry discussions.

For example, in a net metering arrangement, an installed solar power system feeds in to meet a building's electricity requirements, and if the system is generating more than the building uses, the balance feeds into the electric grid, and the meter runs backward. At the end of the year, the total used is subtracted from the total generated, and adjusts the total annual bill. A study by the National Renewable Energy Laboratory estimates that 39 percent of total national electricity sales could be met by rooftop solar installations.[40] However, as Hal Saville, one of the prime movers in SUNWPA (Solar United Network of Western Pennsylvania), explains the solar potential for a specific existing building in any given location can be limited. Only 22 to 27 percent of existing buildings are suitable for solar installations, whether because of roof condition, age of the building, orientation, shading, or other limiting issues.[41] "What can we do for people who want to have a solar system but have property that is too shaded by large trees, or tall buildings? What about people who rent?" Saville asks.[42] One answer lies in creating a community power system. Many places in the country have pursued community-based power systems, either through municipal electric cooperatives, or as community projects.

Each state has different regulatory arrangements, so the renewable energy industry faces the need for unique market assessment for pricing and financing in each state. The tariffs in Pennsylvania are not among those that support aggressive use of renewable sources. The net metering tariff caps how much power a customer can generate beyond current use, and combining customers to net meter from a community installation is not possible here. This limits such options as solar arrays on the rooftops of schools, municipal buildings, or standalone facilities on the edges of roads or fields to serve a number of different customers who share the costs and share the power output. Saville noted, "It is frustrating to see communities with creative and forward-looking ambitions thwarted by regulatory policies mired in the past. As manufacturers, we see many lost opportunities to build solar communities that are resilient for the future. People are trying to do the right thing, but face regulatory frustration."[43]

In the absence of a supportive government policy, a system that responds to customer needs is emerging within the regulatory constraints. A pricing differential could help to add renewable generation in places where it would stabilize or enhance the electric system reliability. The roundtable group discussed the possibility of putting together a pilot demonstration project to take to the Public Utility Commission as a way to begin the process of reforming the tariff system. One solar system advocate said, "If we know that adding solar generation in a specific area would help the reliability and resilience of the grid, we could target our promotion in that area."[44] For instance, if there were an area at the end of a distribution line that had only one feed to serve customers, this area would be more vulnerable to an outage from storm damage to a line. If this area had customer-generated power in excess of the individual customer's load, the power could serve the neighbors in that area when needed. The tariff could reflect the higher value of customer generation in such a situation. Renewable energy providers could target areas of vulnerability for adding customer generation to improve the overall reliability of the system. However, electric utility companies have not yet done updated production cost models or engineering analysis of the system that would be necessary to develop that kind of information.

Infrastructure systems that integrate solar, wind, hydro, and biogas energy for electric service with water and sewer management are not hard to devise from a clean slate. One can imagine zero-net energy/water/waste communities, and the military Net Zero Mission helps to illustrate how this would work.[45] The greater challenge is moving from the existing system of old power plants, existing buildings designed for separate water, electricity, and gas service, and a high cost hurdle for retrofitting these systems.

How can the existing infrastructure be used to build for future needs? A broad analysis of existing buildings, including comprehensive inventories of flat rooftops capable of supporting solar photovoltaic cells, and existing infrastructure such as utility underground corridors, would help to provide an inventory of resources. Underground water and sewer pipes offer the possibility of tapping into geothermal space conditioning by using external heat exchangers on the existing pipes. Currently there is no regulatory mechanism to allow such use.

Arrays of batteries installed as frequency regulation and integration into the grid from diverse distributed generation points offers a new service opportunity for utilities. The location and maintenance of battery arrays for backup power or emergency power already exists in remote locations such as telecommunication facilities or hydroelectric plants. Placing high-performing battery arrays among the distributed generation points can support the growth of customer-based renewable energy systems. Electric vehicles with batteries as storage systems also will have an effect on the demand and storage capability.

Innovations and new technologies are rattling the foundations of the historic concept of public utility services. There are opportunities for new utility services, both for investment in solar installations on customer properties and for offering new kinds of service such as backup systems based on new technologies like compressed air, ice ponds, and direct-current-powered microgrids that may or may not be connected to the central system. The utility system has adapted to large changes in the past, with new tariffs and new regulations emerging to respond. States with competitive markets can play a role in developing a new path forward. As a fully deregulated state, Pennsylvania has the opportunity to develop new ways for the utilities to function within the existing system.

PITTSBURGH PATHWAYS: BUILDING THE FUTURE ON A FOSSIL LEGACY

Pittsburgh sits at the center of the controversy surrounding the transition from fossil fuel combustion to a renewable and sustainable resource base for the economy of the future. Intergenerational obligations play out in the hopes and fears for children and grandchildren of coal miners, and workers in the Marcellus Shale gas fields. Obligations to the international struggle for equity are embedded in choices about what kind of technology and material to export or import. The environmental justice issues of transitioning workers and communities to a sustainable future from a fossil fuel past pose close personal choices. Most critically, people's choices about energy drive the options for survival of the creatures and ecosystems in the path of the extraction process, and suffer the inevitable degradation of living conditions from fossil fuel combustion.

As the old energy systems clash with emerging renewable systems, the human face of the problem becomes contorted with tension. The jobs of the remaining workers in coal, coke, steel, and heavy manufacturing are threatened by increasingly stringent calls for cleaner air, and competition from cheaper natural gas. The new impetus for Marcellus Shale natural gas, touted as a cleaner alternative to coal, produces conflicts over water contamination and health effects as well as land use in residential areas. Even as the evidence of climate change accumulates, the press for aggressive natural gas production from deep shale formations accelerates in Pennsylvania. The prevailing sentiment is that "natural gas will be a bridge fuel to the future."[46] Nevertheless, some communities have banned the practice of hydraulic fracturing, including the City of Pittsburgh.[47]

Conflicting positions about the long-term implications of either continuing on a fossil-fuel-based energy path or shifting rapidly to a renewable energy system play out in the public arena. Political power is unevenly distributed, with federal- and state-level policies more favorable to protecting the status quo and some local initiatives more supportive of changing direction. Pittsburgh mayor Bill Peduto sees government as a force for true progress in sustainability: "We must live in the present with a view toward tomorrow. We are creating an agenda for the future that includes accountability, responsibility, and sustainability. We are creating a culture of governance that recognizes that land, water and sky are shared with the rest of the world and need to be guarded for the future."[48]

Mayor Peduto is fond of referring to Rachel Carson as one of the heroes of Pittsburgh as he points to the potential for new opportunities in the city. "Pittsburgh is a place of hard work, deep values and personal independence balanced with a strong sense of community."[49] Now, individuals, communities, and institutions all over the Pittsburgh area are carving out pathways forward to accommodate renewable energy and sustainable energy systems.

PITTSBURGH: A RESILIENT SOLAR AMERICA CITY

Pittsburgh was named as one of the Department of Energy's twenty-five Solar America Cities in 2007 and as a member of a Rooftop Solar Challenge team in 2011. The city has worked to identify suitable sites for solar power installations, increase solar power on city facilities, and train city employees to install solar systems. After an initial assessment that included energy audits of 325 city buildings, solar hot water installations on firehouses with newer roofs were determined to be the most cost-efficient installations. City public works employees were trained in partnership with Sandia National Laboratories to install and maintain the solar systems. Using a Power Purchase Agreement with local installers who could take advantage of the tax credits, the city has leased rooftops of several buildings for solar power.

Pittsburgh has many vacant properties, including some large brownfields of former industrial property. One site of forty acres has fifteen acres undermined with old coal tunnels that make the land unsuitable for buildings.[50] This area was evaluated as a site for installing a solar field. The city is stabilizing and preparing it for use, with a request for proposals and a power purchase agreement to generate three megawatts of power.

Pittsburgh shares planning and community development efforts with 130 surrounding municipalities in Allegheny County. Pittsburgh is working with its neighbors to adopt a standard zoning permit process to expedite solar installations. A permit for a solar rooftop installation in Pittsburgh is a sixty-five-dollar over-the-counter transaction, while surrounding communities may have varying processes ranging from four weeks to six months at higher costs. Beyond the solar initiatives, Pittsburgh has pursued a more integrated approach that crosses all dimensions of sustainability, integrating the city as a biophilic system in harmony with nature as a goal.

The Millvale Library in Pittsburgh has installed a 9.54-kilowatt solar installation that generates 100 percent of the electric needs for the library. A daily monitoring dashboard allows managers and the community to see the solar energy production on a real-time basis.[51] It is part of Millvale's community vision to become a self-reliant urban solar village. A future project includes plans for a solar cooperative, but this project awaits favorable net metering tariffs that would allow multiple metered customers to connect to the same large solar installation.

"If a community building such as a school or municipal building has enough capacity for multiple users, and people want to buy into the project and share the output, the current system in Pennsylvania does not permit that," said Saville. However, there are three ways to organize community-supported solar projects to allow broader accessibility and participation. In a utility-sponsored project, a utility owns or operates a project that is open to voluntary ratepayer participation. In another option, a special-purpose entity provides a structure in which individual investors join in a business enterprise to develop a community solar project. Finally, in the nonprofit "Buy a Brick" model, donors contribute to a community installation owned by a charitable nonprofit corporation.[52] Many areas of the country are pursuing variations of these three systems to expand the use of solar power.

Fred Kraybill owns one of the largest residential solar installations in the city of Pittsburgh. He and his partners at the Thomas Boulevard Group installed 7.28 kilowatts of roof-mounted photovoltaic collectors in 2012. The building is on a net metering arrangement with Duquesne Light, the local distribution utility. The solar system did offset some of the building's total use, but Kraybill wanted to see how they could do more, so in 2014 a larger array was added in the side garden to bring the total solar system to 18.82-kilowatt capacity. Kraybill also charges his all-electric Chevy Volt from the solar system. He says, "I am driving on sunshine! I think it's so cool that the sunlight that used to hit my roof for 100 years and do nothing but wear out the shingles, is now generating clean electricity via solar panels, and that electricity can run every electrical device in my house and also provides the energy to run my car."[53]

At the 2014 Solar House Tour, on a cloudy October day with drizzling rain, Kraybill's electric meter was still running backward, feeding power beyond his use into the grid. The system was installed by Energy Independent Solutions and was financed using solar tax credits and a net metering tariff. This example is particularly compelling because the building represents so many typical houses of ancient vintage in Pittsburgh and many other older industrial cities. The solar system makes sense economically and offers a highly visible example for others in the community.

CORPORATE PATHWAYS TO RENEWABLE ENERGY USE

Corporate interests have taken renewable energy systems and efficiency improvements to a new level. Motivated not only by the profitability of an energy enterprise but also by a desire to provide stability for the future, many companies are adapting to a renewable energy future. There are numerous examples, more every day. These cases can serve to offer insight into the ethical motivation as well as the business sense for a few innovators.

SCALO SOLAR ROOFING/EDEN HALL CAMPUS

Industry is evolving. Welcome to the future. Welcome to Sunscape is the slogan written on the risers of the stairs to the Scalo Solar company rooftop. The roof, roughly the size of five Olympic-sized swimming pools, is blanketed in solar panels and lush greenery. There is a cluster of thin flexible rectangles, and rows of glistening dark, horizontal tubes. There are five-foot panels perched along the roof in neat rows, configured at different angles facing the sun. Some are mounted over green-roof treatments with drought-tolerant species such as native sedum to test variations in performance against different roof types.

"We wanted to give people a showroom," says Michael Carnahan, the general manager of Scalo Solar. "It's like buying a car. No one wants to buy one without testing it out."[54] And a showroom it is. This is the Scalo Solar Sunscape Demonstration Project, a $1 million endeavor completed in 2011, a few miles west of downtown Pittsburgh. The solar installation on this Scalo Solar Solutions headquarters building saves 78.6 metric tons of carbon dioxide emissions each year.[55]

Everything is arranged to show clients the differences in technologies, efficiency, and placement. "Everything up here is environmentally friendly. These are recycled tires you're walking on. This is a sampling of everything we can do,"[56] Carnahan says. There are four different types of green roofs and three different types of solar panels. But just like a car lot, solar projects come along with sticker shock. "You can fall in love with an electric car, but that feeling can vanish when you see the price tag. The discussion always comes down to finances, so that's where we tend to start," said Carnahan.[57]

Scalo Solar is an outgrowth of Burns and Scalo, a $30 million roofing business that has operated since 1956. Jack Scalo set out to expand into solar roofing when he noticed all the flat black roofs on commercial buildings as he was flying into Pittsburgh. He saw a market for solar rooftops. He envisioned solar panels on the roof embedded in the process of building businesses and homes. Jack Scalo described the company philosophy thus: "This is the way of the future for energy. Putting solar on existing roofs as well as on new roofs helps a company satisfy the moral imperative for sustainability. It is good for the environment, and good for a corporate image. But most of all, solar makes good business sense, for now, and for the future."[58]

"Around 2009 there was a lot of talk about solar, but no action or investment," Carnahan says. Before coming to Scalo, he was skeptical that roofers could be serious about solar energy. But then he saw Scalo's company headquarters roof and he came on board in 2011. "It just made sense, like being on the cutting edge," he said. Carnahan notes that solar can make a big difference in western Pennsylvania because the dominant form of electricity generation comes from coal. "Over its twenty-five-year warrantied lifespan, just one solar panel can produce the same kilowatt-hours as four tons of coal. That makes a huge difference in our environment and our health," he said.[59]

A twenty-year return on investment is an often-repeated number attached to solar projects, and changing minds can be difficult. One of Scalo Solar's major challenges is to demonstrate to engineers and architects that solar projects can have a faster return on investment than typically believed. The traditional equation ignores municipal, state, and federal incentives and subsidies. It also ignores the much lower fuel costs and operating costs. "In reality solar projects can have much shorter returns. We have to explain this to engineers and architects, which is a huge hurdle," Carnahan said.[60]

In general, there are three ways for solar installations to be financed:

1. A private building owner can make the investment directly, applying any federal or state incentives to offset the cost, and displacing all or part of electricity purchased from a utility company to generate savings.
2. A private building owner can contract with a solar company to lease the solar system. In the lease model, a customer will sign a contract with an installer/developer and pay for the solar energy system over a period of years or decades. Solar leases can be structured so that customers pay no upfront costs, pay some of the system cost, or purchase the system before the end of the lease term. Similar leasing structures are commonly used in many other industries, including automobiles and office equipment.
3. A private building owner may enter into a power purchase agreement (PPA) with the solar company. In the PPA model, an installer/developer builds a solar energy system on a customer's property at no cost. The solar energy system offsets the customer's electric utility bill, and the developer sells the power generated to the customer at a fixed rate, typically lower than the local utility. At the end of the PPA contract term, property owners can extend the contract and even buy the solar energy system from the developer.[61]

Scalo Solar owns the solar arrays at Chatham University's Eden Hall Campus. "It's a major investment but it allows us to grow. Being a part of a net-zero campus this side of the Mississippi, it's a big deal," said Scalo. Scalo Solar sells back the solar electricity to the university under a PPA. This arrangement benefits the university because the capital investment for the university was lower, and it benefits Scalo because they can take advantage of the solar investment tax credit not available to the nonprofit university.

The challenges of securing these kinds of projects are enormous. For corporate clients, the standard solar investment would have a return on investment of seven to nine years. But that time frame isn't even in the corporate outlook because most corporations look at a three- to five-year investment horizon. Even for municipalities, it takes a really forward-thinking local government to make that kind of investment. There is also a consideration of whole-life costs. This is especially significant for municipalities or other government entities where the bonding capacity can help by including the cost up front and considering the very low operating costs over the life of the facility. "This concept of life cycle cost analysis has to be integrated into the way everybody does business,"[62] Carnahan says.

Solar power technology is advancing on an almost daily basis as systems are incorporated into building designs. The roof of a building can be designed with solar-collecting shingles, windows that generate electricity as the sun passes through, and walls with collectors mounted as part of the siding. Designing buildings to serve as energy-collecting systems and self-storage units can reduce the profile to zero net energy requirements. "As we go forward, if you have a roof, it is going to have solar on it!" says Carnahan.[63]

WINDSTAX: RENEWABLE ENERGY FOR INDUSTRY

Ron Gdovic, owner and founder of WindStax, made a deliberate decision to pursue an unconventional approach to harnessing wind power. Armed with a mechanical engineering degree, the tenacity of a serial entrepreneur, and welding skills honed in his father's sheet metal shop, he intends to make small-scale wind power ubiquitous.

Why wind energy? Wind energy is a domestically produced, pollution-free, renewable energy source that can be built on existing farms and ranches.[64] Wind power is scalable. Large wind farms are more efficient, require less water, and incur fewer operating expenses than conventional power plants. A single large wind turbine requires just a quarter acre of land and units can be distributed throughout existing farms and ranches and provide annual royalties of over two thousand dollars.[65]

However, harnessing the power of the wind can be difficult. Wind is intermittent and wind speed varies from place to place. Areas with wind power at class 3 (wind speeds of up to 15.7 miles per hour, which is suitable for conventional wind energy production) are dispersed throughout coastal areas, the Great Plains, portions of Alaska, and mountain ridges and summits throughout the Rockies and Appalachians.[66] Areas with the strongest winds, like western Montana and the Dakotas, are far from large energy-using metropolitan areas, and transmitting energy to more densely populated areas would decrease efficiency.[67] There are also concerns about the aesthetics of large conventional windmills, which can tower over three hundred feet in the air; concerns about the safety of windmills, which may pose a threat to birds and bats;[68] and concerns about the noise, which is particularly troublesome in previously quiet areas like Vinalhaven, Maine.[69]

But many areas of the country have lower-velocity winds that blow on a regular basis. Ron Gdovic founded WindStax, his eleventh company, in 2012 and hopes to bring wind energy everywhere from remote farms to urban housing developments and industrial facilities.[70] His business manufacturing twenty- and forty-foot vertical wind turbines in Pittsburgh's Strip District is working hard to keep up with demand.[71] Facing a turbine design that hasn't changed much since the 1600s, when windmills were used to power sawmills in New York City,[72] and a long list of constraints from low wind speed to concerns about noise, Gdovic sought a creative strategy. He started with a decades-old sketch based on a vertical design by Finnish engineer Sigurd Johannes Savonius. He took a few years to modify the Savonius design[73] and to build prototypes for "an affordable, reliable, and effective wind turbine that works as well in urban areas as they do anywhere."[74] Gdovic's patented design is a smaller, quieter, vertical-axis turbine. It operates at lower wind speeds, where the competition can't operate, skirts the bird issue, and is a self-contained plug-and-play unit, so structural engineers or other experts are not needed to install the unit. Designed in the spirit of manufacturing with the complete life cycle supply and waste stream in mind, WindStax turbines utilize low-volatile organic componds as finishes and incorporate retired utility poles as a building material. The turbines are designed to be recycled at the end of their fifteen-to-twenty-year life span.

With hand-squeezed rivets, personalized colors, and a structure that is beautiful as well as functional, Gdovic creates a value-added product that is difficult to replicate. According to Gdovic, his customers appreciate this attention to detail and careful craftsmanship. Customers often name their turbines and "do the sales pitch for us" as they show off their turbines to customers, friends, and neighbors.

Gdovic finds it especially gratifying to build something tangible and to watch his customers' faces light up when the turbines are installed.[75] Maybe it's the Pittsburgh working-class ethic shining through. In contributing to his hometown of Pittsburgh, Gdovic has placed a priority on hiring employees that were underemployed or unemployed and has recruited through CareerLink, which places unemployed union workers in new jobs.[76] He also located his company in the heart of the city of Pittsburgh and is using infrastructure that originally housed the birthplace of the Aluminum Company of America, ALCOA. The result is a company where employees are excited and motivated to come to work, where customers are product ambassadors, and where demand for WindStax turbines exceeds supply. Customers include Epic Metals, with six WindStax turbines installed at its facility along the Monongahela River in Braddock, and PITT OHIO trucking company, which has installed WindStax turbines as well as solar photovoltaic arrays in its operation in Illinois and New Jersey as well as Pittsburgh. Gdovic says, "I have young daughters, and this technology will give a cleaner energy future for them. It is a legacy to be proud of."[77]

The City of Pittsburgh and its mayor Bill Peduto are also interested to see how clean energy startups like WindStax can grow Pittsburgh's economy.[78] Perhaps in the future there will be a plaque at 3200 Smallman Street that notes this location as not only the birthplace of the aluminum industry in the United States but also the birthplace of small-scale wind energy.[79]

CONCLUSION

Pittsburgh is creating a path toward a sustainable future amid the stronghold of fossil and nuclear interests. Through projects large and small, and with public support, a shared vision for a future that offers resilience in the face of change is emerging. Active discourse fueled by the many university students in Pittsburgh brings issues of ethics and justice forward. Even as demonstrations for divestment from fossil fuel industries fill the streets, major companies such as PNC Bank are making investments in zero-net-energy buildings.

Intergenerational ethics issues are addressed by pursuing an energy system that does not leave a legacy of long-lived carbon dioxide in the atmosphere, wastes that remain hazardous for thousands of years such as spent fuel cores from nuclear power reactors, or cause unknown effects of highly saline and chemical-laden fluids pumped into the ground for hydraulic fracturing. Instead of exporting fossil fuels to needy countries, especially in emerging economies, Pittsburgh has the potential to export the technologies and systems for making renewable energy communities work. And in a city closely identified with smoke and pollution from power and industry, renewable energy systems present sustainable solutions. With deep roots in unions and workers' rights, the transition from fossil industry work to sustainable energy system jobs can lead a time of sustainable prosperity.

Renewable energy, naturally distributed everywhere, combined with modern communication technology creates new possibilities to colocate the use of energy with its generation. The central station, fossil-fuel-powered energy system of the Victorian age, with its waste and combustion emissions, is being replaced with distributed energy systems suited to the need for a low-carbon emission future. People begin to feel empowered to make a difference in their own homes, communities, and businesses one step at a time. The energy system transformation is underway.

Kretschmann Family Organic Farms grows certified organic fruits and vegetables on their eighty acre farm.

KRETSCHMANN FAMILY ORGANIC FARM

5

RESTORING FERTILE GROUND

The food we eat, the food that nourishes us, is a gift from the earth, from the sun, from millions of years of evolution.

VANDANA SHIVA

FOR THE MODERN American, food is something best dealt with quickly, with little fuss. And although three-quarters of Americans are eating meals at home each night, the number of meals prepared at home continues to decline, with almost half of the food served from preprepared or prepackaged meals at fast food stores or restaurant and supermarket takeouts.[1]

Americans, at the top of the food chain and with the arrogance of a dominant culture, expect any food to be available at any time. One-third of American cities have farms within one hundred miles, but food sold in stores frequently travels thousands of miles. Fruits and vegetables go from local farms to distant sorting, packaging, and processing centers to be reshipped, sometimes back to the area of origin. People assume that food is abundant, healthy, and safe, and available all year round. In many cases the food system itself compromises quality and nutrition in exchange for convenience in shipping or processing.

Many people have little idea of where food originates. The fact that some vegetarian soy products caused Papua New Guinea ecosystems to collapse as permaculture forests turn to soya fields is not posted on the label. Shoppers have no way to know that the fashionable specialty quinoa grain is sending the price of the native crop in Chile out of range for local people there. Coffee that is a fair trade product, with just compensation to the grower, competes with the cheapest commercial brand on the shelf.

Global trade agreements and global advertising connect the world food market in a vast global supply network, and advertising for industrial food businesses manipulates food trends for massive profits. For example, the American meat-heavy diet has been exported as a sign of affluence all over the world. Ninety-five percent of Americans eat a third to a half a pound of meat every day. In Japan, McDonald's fast food stores thrive serving hamburgers to a culture that thrived for centuries on rice, vegetables, and fish. Americans at the beginning of the twenty-first century are consuming more food and several hundred more calories per person per day than did their counterparts in the late 1950s, when per capita calorie consumption was at the lowest level in the last century. In 2000, total annual meat consumption (red meat, poultry, and fish) reached 195 pounds per person—57 pounds above average annual consumption in the 1950s.[2]

This heavily meat-dependent pattern of food choice raises significant ethical issues of international justice as well as issues of rights of living things to exist. The modern propensity to eat more meat and more processed foods has a resource burden as well as a nutritional burden. Vegetables, fruits, and nuts require about twice the energy and water to produce as grains, and meat and dairy products require about four times the water and energy resources as grains. Overall, it takes ten times more resources to produce a pound of beef than of grain. The agricultural industry produces four categories of products:

- Vegetables, grains, fruits, nuts and seeds for direct human consumption
- Animals and animal products for direct human consumption
- Grains and feed for animals
- Plant material for fiber and fuel production

TABLE 5.1. RESOURCES NEEDED FOR DIFFERENT AGRICULTURAL PRODUCTS

FOOD PRODUCT	ENERGY FOR PRODUCTION (BTU/POUND)*	WATER USED (GALLONS/POUND)**
Grains	5,750	154
Vegetables	11,600	257
Fruits	12,750	420
Nuts	13,600	368
Dairy	16,400	381
Meat	22,500	1,847
Eggs	23,500	573

Notes: * Michael E. Webber, "More Efficient Foods, Less Waste," *Scientific American*, December 29, 2011, www.scientificamerican.com/article/webber-more-efficient-foods-less-waste (accessed July 10, 2014).

** M. M. Mekonnen and A. Y. Hoekstra, *The Green, Blue, and Grey Water Footprint of Crops and Derived Products*, vol. 1, main report, UNESCO-IHE Institute for Water Education (Netherlands, 2010). Note that the data has been converted from liters/kg to gallons/pound.

The share of resources needed for production varies greatly among these types of agricultural products. Table 5.1 illustrates the variance in energy and water resources required to produce the main categories of foods.

These differences in the resource requirements for producing different types of foods have ethical implications as the press of population on arable land increases globally. This is especially significant because the cropland and resources used for growing and feeding animals for food (26 percent of developed land) far exceeds the agricultural land used for growing food eaten by people directly (19 percent of developed land).[3] Land pressure for agricultural use pushes into fragile areas, and areas are converted to agriculture with a great cost in loss of biodiversity.

The United Nations goal for food security is for "all people, at all times, [to] have physical, social and economic access to sufficient, safe and nutritious food to meet their dietary needs and food preferences for an active and healthy life."[4] To achieve food security, nutritious and wholesome food must be accessible to people at a reasonable cost. Food must be delivered in a form that people know how to use and prepare, and the supply must be stable in all seasons, even in crisis conditions and emergencies. These conditions of food security apply all over the world, regardless of economic conditions.

How we gather, prepare, share, and store food defines our cultures more distinctly than any attribute besides language. Ways of obtaining food also define the relationship of people to the environment because no other aspect of life is so critically bound to the living earth. As the global population hurtles past seven billion, global food productivity is shrinking. Even as the demand for food to nourish more people increases, the degradation of agricultural conditions is also increasing from the pressure of human development. Poverty and hunger loom. Restoring a balance between what the earth can provide and what people use is critical for our survival as a species. Our health and well-being are only as secure as the health of the living earth.

In meeting the food sufficiency for the next century, three great needs must be balanced:

- To feed more than nine billion people by 2050
- To provide economic opportunities to alleviate poverty
- To reduce pressure on the environment[5]

Prevailing practices for food production are ill-suited to meet these goals. "Society is undermining the ecological foundation of its own food system," concludes Joseph Alcamo, chair of the United Nations Environment Programme Synthesis Study on Avoiding Future Famines.[6] The security of the food system depends on the health of ecological foundations, which are being threatened in two ways: by undercutting the natural conditions needed for agriculture—freshwater, soil formation, biodiversity, pollination, and many other services; and by human-produced side effects that compromise food production, such as greenhouse gases, pollution of the air and water, and biologically toxic effects of chemicals. An ethic of preserving the land as well as justice in food availability and access adds urgency and global context for solutions to balancing food, water, and energy.

Challenges to sustainable agricultural practices are also social and cultural.

To produce food sustainably, the needs of people as well as the needs of the land itself must be addressed. This is more difficult because, while the physical challenges in agriculture can be addressed by changing the techniques and practices used for production, meeting needs of people requires attention to the cultural values across diverse global interests.

Agriculture stands at a crossroads. The current path clearly leads to infertile farmland, destruction of essential ecosystems, and broader accumulation of toxic contaminants. However, there is reason to hope for a better future because the remedies for this situation lie readily at hand. Reversing the trend from larger, more mechanized, and chemical-dependent monoculture farms to smaller farms with cultivation practices based on agroecology can restore the fertility of the land, increase food security on a global scale, and reduce the impact of human food production on the biodiversity of species. A pathway toward a sustainable food supply is emerging based on the preservation of fertile ground through improved agricultural practices in small gardens as well as large commercial farms.

UNSUSTAINABLE PRACTICES OF INDUSTRIAL AGRICULTURE

William M. Muir, a professor of animal sciences at Purdue University, claims, "If there ever was a major improvement in human life in the last millennium, it may well have been the green revolution of the twentieth century. The massive use of fertilizers, pesticides and improved seeds and livestock brought an enormous increase in agricultural production, more than enough to feed the population."[7] But this improvement has come at a cost. The principle of precaution in applying new technology that affects natural systems of the world has been violated on a grand scale in industrial agricultural practices. The modern industrial agriculture system creates a legacy of degraded soil and chemically contaminated land and water, as well as ecosystem disturbances that will persist for generations.

The widespread use of large-scale monocultures is now the primary way of growing commodity products such as corn, soy, and wheat. This means that large areas are planted with the same crop, allowing farmers to use one kind of cultivation technique and machinery. While seemingly more efficient, the cultivation of monoculture crops depends heavily on herbicides to control weeds, and pesticides to control insects. Much of the 915 million acres devoted to agriculture depends on highly mechanized processes and chemical fertilizers, herbicides, and pesticides.[8]

Increasingly, growing commodity crops such as soybeans, corn, rapeseed, and cotton, is based on using genetically modified seeds to resist insect pests.[9] In this system, the commercial crop has been genetically modified to be unaffected by proprietary herbicides and pesticides. Fields planted with this crop are sprayed from airplanes to apply massive quantities of these herbicides and pesticides that, though harmless to the crops, are broadly destructive of non-genetically-modified species in the area. This has had a devastating effect on biodiversity both on the farm and in off-farm areas such as hedgerows and byways, often the habitat for wildflowers and sources of food for pollinators and wildlife. Thus, this pest- and weed-control practice even further enforces the monocultural nature of the cultivated lands.

As a consequence of this type of farming, the farmlands and the surrounding areas suffer from a severe lack of ecosystem-level biodiversity—there is no interspecies interaction, no give-and-take of nutrient cycling. Many of the naturally occurring supporting ecosystem services have been eliminated. As might be expected, this has a number of serious negative consequences. Because the crops are grown as monocultures, the planted crop uses up all of the nutrients in the soil that it requires, and because the same crops are planted on the same land over and over again, those nutrients are stripped from the soil and never replenished. The natural fertility of the soil is destroyed. Moreover, even if the soil were fertile, the crop would still be inefficient—because the crops are cultivated, rather than native, species, the plant is already ill-adapted to the soil in which it is planted. Due to the combination of all of these factors, massive quantities of nitrogen-rich, petrochemically derived fertilizers, as well as massive amounts of irrigation, are required in order to maintain consistent crop production.

The cycle of pesticide, herbicide, and fertilizer overuse in farmland has several unintended consequences. In the first place, as noted above, pesticides that are based on synthetic chemicals often indiscriminately kill beneficial organisms, such as pollinators and insect parasites and predators of pests. Productivity has plateaued as living systems are saturated by chemical burdens. Agricultural practices also contribute to water supply contamination and

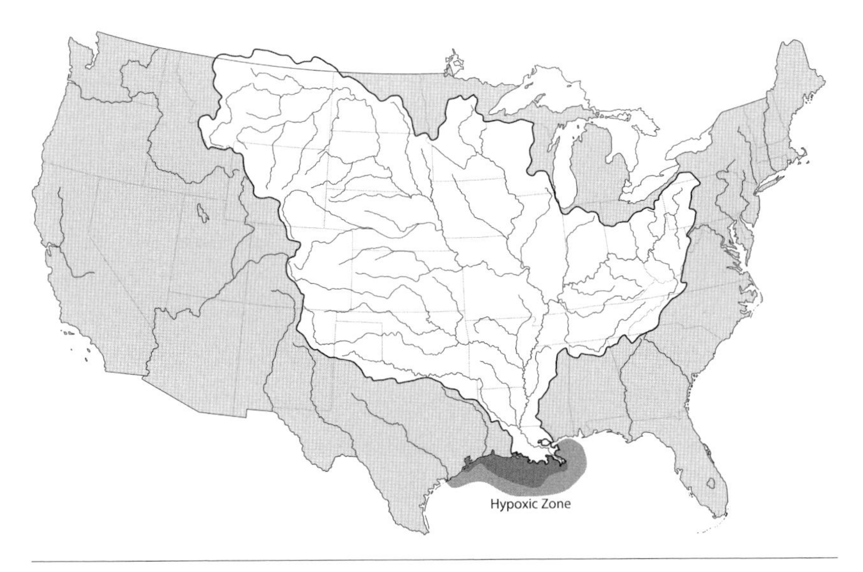

The Mississippi River drainage basin extends over more than 1,245,000 square miles, including all or parts of thirty-two US states and two Canadian provinces, ultimately emptying into the Gulf of Mexico.
MAP DRAWN BY BILL NELSON

degradation from farm runoff and the leaching of fertilizer, pesticides, and herbicides into waterways. A visible example of this effect of the excessive use of chemicals in commercial agriculture can be seen in the map above, which shows the Mississippi River drainage area and the dead zone in the Gulf of Mexico affected by the runoff from farmland. The dead zone area is where water is depleted of oxygen from an overburden of nitrogen—the nutrient added by chemical fertilizer—that flows from farm fields. The area of the dead zone varies seasonally, averaging 5,500 square miles—an area the size of Connecticut and Rhode Island combined.[10]

In addition, the repeated use of pesticides can induce resistance in some pest species. Since 1945, an estimated one thousand species of agricultural pests have developed resistance to at least one pesticide,[11] and the US Department of Agriculture has identified at least 451 major agricultural pests that have developed resistance to multiple pesticides.[12] Even genetically modified crops encounter resistant pests within three to five growing cycles. This has resulted in an ongoing "pesticide treadmill," wherein increasingly toxic materials are developed in an "arms race" to stay ahead of adaptive insects and "weeds."

Another serious consequence of seeds hybridized for compatibility with chemical pesticides is that the genetic diversity in crops has been drastically reduced. Industrial monoculture farming means that a single stock of crop that has been modified to be resistant to the herbicide/pesticide is perpetuated to the exclusion of any other strains. In American farming production, 97 percent of vegetable varieties available in 1900 are now extinct. From 1903 to 1983 over six thousand varieties of fruits and vegetables were lost, including 86 percent of tomato varieties, 93 percent of lettuce varieties, 90 percent of field corn varieties, 96 percent of sweet corn varieties; and 86 percent of apple varieties.[13] Because all individual plants are genetically the same and lack the resilience to adapt to changes, this lack of genetic diversity renders crops even *more* susceptible to disease and pests. And it also means that crops are more vulnerable to variations in growing conditions such as drought—variations that are likely to increase as a result of global climate change.

ERODING NATURAL CONDITIONS FOR AGRICULTURE

According to Achim Steiner, UN Undersecretary General and Executive Director of the United Nations Environment Programme, "Up to 25% of the world food production may become 'lost' during this century as a result of climate change, water scarcity, invasive pests and land degradation."[14] These effects apply in the United States as much as they do globally.

The American Farmland Trust estimates that 125 acres of fertile farmland is lost per hour across the United States. There has been an 8 percent decline in the number of acres in farms over the last twenty years in the United States. In 1990 there were almost 987 million acres in farms in the United States; that number was reduced to just under 943 million acres by 2000, and then reduced to 914 million acres in 2012.[15] This is partly a result of the incursion of development into former wilderness and farmland. Population growth as well as increasing affluence presses development into areas otherwise available for growing food—grasslands, forests, and wetlands. Development pressure comes from sale of farms for buildings as suburbs continue to expand. Energy extraction industries also decrease arable land when open farmland converts to industrial uses such as hydraulic fracturing for natural gas that lies beneath much of the most fertile ground in America.[16] Compounding the losses sustained due to development is the loss of arable land through degradation. Areas under intense monoculture cultivation or grazing for long periods of time experience degradation from nutrient depletion as well a material loss of topsoil. In the Great Plains, for example, topsoil, measured at fourteen to twenty-four inches in 1900, has fallen to four to six inches in the most intensely farmed areas today.[17]

Climate change affects the productivity of agriculture in many ways. For example, changes in climate conditions have reduced the productivity of fisheries as the increased acidity of the oceans has affected aquatic habitat. In the Pacific Northwest, oyster seed must be set in controlled nurseries fed by seawater to allow the shells to form before being set into the open water; however, in recent years, oyster seed has been scarce because the conditions for initial shell formation in the seawater are too acidic.[18]

As the average temperature of the planet rises, zones of cultivation shift, making some areas into deserts and extending the growing seasons in other places. Shifts in habitat, growing seasons, and patterns of rainfall contribute to the instability of food supplies. Warming global temperatures affect the hydrological cycle, converting areas of regular rainfall to areas of prolonged drought and reducing the annual average snowmelt contribution to the flow of rivers and freshwater streams. More frequent and unpredictable severe weather events also disrupt regular agricultural cycles. Too much rain at the wrong time can destroy crops, just as failure of regular rainfall can parch the ground and make it susceptible to dust storms.

Industrialized agriculture aggravates the problem of climate change as well as being adversely affected by the changes. Globally, agriculture contributes about 25 percent of the greenhouse gases that cause climate change.[19] These emissions come from the methane emitted by agriculture for livestock and manure pools, from nitrous oxide coming from fertilizer, from carbon dioxide emitted by tractors and other farm machinery, and in the production and application of fertilizer, herbicides, and pesticides. The agricultural production of ethanol and biomass used for fuels also exacerbates greenhouse gas emissions that contribute to global warming.

Proponents of biofuels argue that using them to reduce the combustion of fossil liquid fuels such as gasoline and diesel can control greenhouse gases and air pollution. In some states, mandatory ethanol additives to gasoline are required as part of this agenda, but also to add a subsidy to growers of biofuels.[20] But some studies show that making fuel from food crops has a net negative energy balance—that means it takes more fuel and resources to develop it than is recovered when it is used.

Researchers have found that the expansion of biofuels from food crops in developed countries affects developing countries by increasing food prices, reducing the products available for human nutrition, and increasing malnourishment.[21] Faced with shrinking farmlands and competing demands on resources, we face pressing questions about the value of resource allocation for food vs. energy production. When food crops are directed to use for energy production, such as corn for making ethanol as a fuel additive, the global food markets for grains are affected. This is especially onerous when it is technically possible and effective to create biofuels from waste material and nonfood sources in developed countries where they are used.[22]

In areas of California and Texas, the competition over water use between energy production and irrigation has created serious controversy.[23] As the new boom in deep gas extraction by hydraulic fracturing has spread across the country, the conflict between food and energy has emerged because the process uses so much water. Hydraulic gas extraction from deep shale formations use up to five hundred thousand gallons per fracking event, and each well may be fracked several times.

America, along with China, India, West Asia, and North Africa, is among the areas where more than 35 percent of groundwater withdrawal exceeds recharge volumes.[24] This means that water is withdrawn from ground sources at a faster rate than it is replaced from rainfall and stream infiltration into the soil. When groundwater is scarce, irrigation, livestock, industrial uses, and domestic supplies all compete for water drawn from aquifers. When agricultural fields are watered continuously from underground aquifers, the salts and minerals from the groundwater can build up in the soil. Lower groundwater tables and increased salinity have already become serious problems, especially in western states.

Water scarcity results in part from changes in climate, especially as high mountain ice packs melt and do not reform seasonally, as they have in the past, to replenish the water supplies of land in the valleys. The conditions in California during 2014 and 2015 highlight this situation, as growers struggled to keep crops growing in a prolonged drought period. Yet the water scarcity issues, in America at least, are mostly manmade. The greatest water problems facing the United States are not lack of infrastructure, but inefficient use, inappropriate water allocations, water pollution, and ecological destruction.[25]

In Pennsylvania, the American Farmland Trust shows the colocation of the Marcellus Shale gas field directly aligned with the most fertile areas of western Pennsylvania, New York, Ohio, and part of West Virginia.[26] This area is particularly significant as agricultural land because climate-change mapping indicates that this is among the areas where rainfall and relatively temperate seasons will continue for a longer time. These are farmlands that form a critical food cultivation resource. However, attempts to preserve agricultural uses and protect watersheds meet two hundred years of property law and heavy legal precedent in favor of energy resource development. There are few legal precedents that claim priority for natural systems or the rights of future generations.

The decisions about choosing to preserve fertile ground for the future as opposed to extracting resources for profits in the next twenty years really rest on determinations of intergenerational fairness. In Pennsylvania, legal debate and interpretation of land use disputes is revolving around the Pennsylvania constitution. Article I, Section 27, of the Pennsylvania constitution states, "The people have a right to clean air, pure water, and to the preservation of the natural, scenic, historic and esthetic values of the environment. Pennsylvania's public natural resources are the common property of all the people, including generations yet to come. As trustee of these resources, the Commonwealth shall conserve and maintain them for the benefit of all the people." Evaluating land use options in this context may place a greater weight on the needs of future generations over short-term economic benefits.

MOVING TOWARD REGENERATIVE AGRICULTURE

Providing a system for making nutritious food available to all economic segments of society is a fundamental criterion for social justice. For a sustainable future, the natural systems that support humanity, especially food systems, must be preserved. Our sustainable future will be built on a food system that integrates human activity into an ecosystem of practices that restore the living earth. The ethical considerations must extend to considering the fertile ground as part of the legacy to the next generation. For that to happen, the living soil with its myriad life forms must be regenerated, not degraded, by agricultural cycles.

The advance of technology throughout the twentieth century has brought tremendous changes in the practice of agriculture. In many ways, using machines and chemicals to produce crops on a large scale has liberated people from subsistence farming, allowing the pursuit of other enterprises for the good of mankind. Nevertheless, commercial agriculture depends on manual labor, especially seasonal workers, many migrating from Mexico and South America to work as itinerant labor.[27] According to the US Department of Agriculture, "hired farmworkers make up less than 1 percent of all U.S. wage and salary workers, but play an essential role in U.S. agriculture. In 2012, some 1.1 million hired farmworkers on average were employed on U.S. farms. . . . Of those crop workers surveyed between 2007 and 2009, 71 percent were foreign-born (67 percent in Mexico and 4 percent elsewhere). Forty-eight percent of crop workers surveyed indicated that they were not legally authorized to work in the United States, down slightly from the peak of 54 percent in 1999–2001."[28] Industrial-scale farming has reduced the farmworker to a unit of production where cost control is the predominant measure. The prevailing system raises ethical implications of farming with workers willing to accept low wages out of fear of deportation or reprisals.

At this stage of civilization, increasing agricultural productivity means that practices that deplete and degrade the land must stop. Instead, practices for farming that respect the laws of nature and regenerate the fertility of the land need to be encouraged and supported. Agriculture must move from the age of technology to the age of biology to make further progress, including consideration for the dignity and well-being of the farmworkers.

The social and cultural challenges to provide food for a growing global population within constrained resources means that the ethical issues must be a larger part of the decision process. Understanding the options available can help to clarify choices for moving the agricultural and food supply systems to a more sustainable pathway. As with energy, the choices are not dependent on technology or great breakthrough adaptations. Rather, solutions can come from enhancing productivity within the ecosystems that support growing healthy food. Supportive policies can restore the fertility of the soil, and the soul of the land, through ecology-based agricultural practices.

ORGANIC FARMING

Who has the right to decide for the countless legions of people who were not consulted that the supreme value is a world without insects, even though it be also a sterile world, un-graced by the curving wing of a bird in flight?

RACHEL CARSON

Over five billion pounds of pesticides are used in US farms each year—about 25 percent of pesticide use worldwide.[29] Many of the pesticides and herbicides used on fields find their way into foods that retain residues of chemicals used in their cultivation. The US Centers for Disease Control reports forty-five thousand human poisonings from pesticide exposure annually, three thousand of which are serious enough to require hospitalizations.[30] The majority of these incidents involve children.

Organic crop management significantly reduces residues of chemicals found in food products. A recent study comparing commercially grown, organic, and integrated pest management (IPM) agricultural products found residues of pesticides in all samples, but at vastly differing frequencies. Organic farming makes minimal use of off-farm inputs and relies on management practices that restore, maintain, and enhance ecological harmony. Integrated Pest Management (IPM) is a system that uses similar practices, but may use chemical intervention if monitoring indicates the need for limited applications. Commercial agricultural practices vary in the use of chemical herbicides, pesticides, and fertilizers, with applications ranging from limited use to aerial spraying and cultivating genetically modified organisms that tolerate these chemicals.[31] Overall, across eight fruits and twelve vegetable crops, 73 percent of conventionally grown samples had pesticide residues.[32] For five crops (apples, peaches, pears, strawberries, and celery) more than 90 percent of samples had pesticide residues; and pesticide residues were found in 47 percent of the IPM samples.[33] Organically grown samples consistently had far smaller percentages, with residues in 23 percent of samples.[34] Most of the residues in organic foods (and some of the residues in conventional foods as well) can readily be explained as the unavoidable results of environmental contamination by past pesticide use, or by "drift" (sprays blown in from adjacent nonorganic farms). Some foods sold as organic may also be mislabeled, either because of fraud or because of lapses in maintaining the identity of foods as they move from the farm to the consumer.

Forty percent of the 2.3 billion acres of US land is devoted to agriculture, with less than 1 percent of the total registered as certified organic farm or pasture.[35] But the practice of organic farming is growing rapidly, and many smaller farms practice organic and sustainable agriculture without undertaking the costly certification process.[36] Moving food production systems from conventional chemical-intensive practice to organic management can increase productivity, reduce contamination, and conserve both water and energy.

A Rodale Institute Farming Systems Trial led by David Pimentel compared the effects of organic farming practices to conventional farming practices over the course of twenty-two years.[37] The study found that farming organically induced less soil erosion, conserved more water, maintained soil quality, and conserved more biological resources compared to conventional, nonorganic farming practices.[38] Organic grain production systems increased soil carbon 15 to 28 percent, while soil nitrogen in the organic systems increased 8 to 15 percent, whereas the conventional system showed no significant increases in either soil carbon or nitrogen in the same time period. Drought and disease resistance were also improved in the organically managed fields. The Rodale Institute study found that organic farming systems require about 15 percent more labor,[39] but "the higher prices that organic foods command in the marketplace still make the net economic return per acre either equal to or higher than that of conventionally produced crops."[40]

Rather than consuming fossil energy to generate greenhouse gases, the soil of an organic farm can sequester carbon dioxide at the rate of about one thousand pounds of carbon per acre-foot of soil each year.[41] Thus, organic agriculture binds carbon into the soil and removes it from the atmosphere, performing functions in the reverse of combustion. A study published in 2015 found that "replacing chemical fertilizer with organic manure decreased greenhouse gas emissions"—carbon dioxide and nitrogen oxide—"which reversed the agriculture ecosystem from a carbon source . . . to a carbon sink."[42] Forming a carbon sink in the ground not only restores fertility on a sustainable basis but also helps to control the further accumulation of carbon dioxide in the air.

EVALUATE THE UNINTENDED CONSEQUENCES OF GMOS

Long before people knew for sure what genetic material was made of and the mechanisms by which inheritance worked at the biochemical level, farmers practiced animal husbandry and plant hybridization to improve the quality of foods. Folklore and traditional stories are full of the search for useful medicinal herbs. And from earliest recorded times, people have managed their environment to improve their living situations and their opportunities for survival in a world that could sometimes be hostile.

In the twentieth century, increased knowledge of how the genetic system works has opened new possibilities. Genetic engineering, the capability to manipulate genetic information outside of organisms, has opened intoxicating vistas of experimentation and exploration. Combinations not only within species but across species have become possible. Scientists can add functions and attributes to foods, or add properties to animals, and can manipulate entire environments on a grand scale. There is no question that genetic modification techniques have made great achievements, such as the synthesis of human insulin through genetically modified *E. coli*, or the addition of nutrient enrichment to food crops. But what is the potential for disruption? What is the effect on biodiversity? What happens to the natural ecosystem services as a result of unconstrained or accidental release of genetically modified organisms (GMOs)? What happens ultimately to people?

Without restrictions on broad dissemination of GMO species, unknown interactions may occur, possibly to damage native seed stocks, possibly to weaken the diversity of natural defenses. The GMO/pesticide-herbicide tango has not yet played out over a whole generation and we do not know where the dance will end. Rachel Carson expressed her concern about the increasingly industrialized agricultural system: "We are subjecting whole populations to exposure to chemicals which animal experiments have proved to be extremely poisonous and in many cases cumulative in their effect. These exposures now begin at or before birth. No one knows what the result will be, because we have no previous experience to guide us."[43] Though she made these remarks in the context of some chemicals no longer used in agriculture, such as DDT, the same principle of being concerned about the cumulative effects and lifelong exposures continues.

Our laws have not kept pace with these innovations and fail to consider the potential consequences such innovations may have on natural systems. The regulatory approval process that reviews new chemicals for use as pesticides, herbicides, or fertilizers does not cover the GMO system in the same way. Some GMO creations have been considered "pharmaceuticals" or veterinary products, such as disease-resistant GMO salmon. There is an increasing demand for clear labeling of GMO foods, but this is complicated by the use of GMO products in many processed foods as component substances such as starch, high-fructose corn syrup, or feed for meat-producing animals. Industries that make both GMO seeds and the pesticides and herbicides they are paired with vehemently oppose labeling, and so far, have successfully suppressed federal standards.[44] In the instance of GMO crops and use of pesticides and herbicides, the question Rachel Carson raised is more urgent than ever: "I recommend that you ask yourself—Who speaks? And Why?"[45]

PERMACULTURE

Proponents of monoculture, with its heavy chemical dependence, defend the practice as efficient and necessary; however, practitioners of permaculture have a totally different point of view. A permaculture approach to food cultivates an agricultural landscape that includes care of the natural ecosystem and care for human cultural and spiritual needs as well as providing sustainable harvests. The concept emerged in the 1970s through a collaboration between Bill Mollison and David Holmgren, who define permaculture as "consciously designed landscapes which mimic the patterns and relationships found in nature, while yielding an abundance of food, fiber and energy for provision of local needs."[46]

Permaculture unites the concepts of permanently sustainable agricultural practices with attention to human cultural values. It is more a system of thinking than a prescription for specific cultivation methods, and places high priority on restoring and building the fertility of the soil then relying on the natural processes of the land to produce goods for human use. Permaculture incorporates an ethic of placing human actions in producing and gathering food within a framework that maintains biodiversity and preserves the integrity of natural ecosystem services.

While food production is one part of permaculture, its twelve principles define an approach to how humans use the earth's resources, and constitutes a cultural shift from an extractive and exploitive approach to a restorative and renewable approach.[47] The process is more labor intensive, but has the benefit of restoring fertility to the soil, rather than depleting it, and supports biodiversity both on the farmed land and in the surrounding areas. The permaculture approach to growing food incorporates an ethic of care for the land to improve the fertility of the soil for future generations. Whether adopted in a large community with multiple farms, or in a single garden, using land in ways that maintain or restore the life of soil organisms—microbial and fungal and insect worms, nematodes—permaculture builds the fertility of the soil transforming it from a sterile substrate to a life-giving mechanism for transferring minerals and nutrients to growing plants. Homberg writes, "Planting trees and other perennial vegetation to restore the health of the land, without the need to gain economic benefit, has been a central activity of the permaculture and broader Landcare movements. In its purest form, conservation of indigenous ecosystems in small reserves implies a commitment to manage in perpetuity, systems otherwise unable to survive broad ecological changes."[48] As the climate changes from human actions, using permaculture as an integrating land ethic can build resilience and a more sustainable pathway for human interaction with the earth.

REDUCE FOOD WASTE

Americans waste 40 percent of the food produced for human consumption.[49] Food is wasted at all stages of the production chain, but the largest point of waste is consumer loss from discarding food that is not eaten. The average American wastes twenty pounds of food each month.[50] Thousands of pounds of food are wasted for no reason related to food quality. If the package that contains fresh produce is damaged, the whole thing is tossed in the dumpster. If the outer leaves of a cabbage are bruised in shipment, it is tossed. If the shelf life date is past, food is discarded, even if it is not spoiled. Shoppers expect "perfect" produce and attractive packaging and displays. The average American consumer wastes ten times as much food as someone in Southeast Asia[51] and up to 50 percent more than Americans wasted in the 1970s.[52]

At all stages of the food supply chain preventable losses occur. In food production at the farm level, 20 percent of the food produced is wasted from not being harvested, or from being lost between harvest and sale.[53] The reasons for leaving food unharvested in fields and orchards vary greatly. There may be crop damage from weather or disease, market conditions may reduce demand, or labor shortages may impede harvest. Approximately 7 percent of all planted fields are not harvested in the United States each year.[54] In the process of moving food from farms to the point of sale much is lost in the transportation, packaging, and distribution processes. Produce is culled before shipping.[55] Culled produce may be sent in bulk to processing facilities, but this is limited by proximity and quantity for bulk quantities. Distribution and retail losses result in 10 percent of food wasted.[56] At the point of sale, items that have been bruised or deteriorated during transport are discarded before the product is placed on sale for customers. And finally, Americans tend to buy more food than they consume—a third of seafood, 28 percent of fruits and vegetables, 27 percent of grain products, and 12 percent of meat purchased are not consumed and end up in the garbage.[57]

Even in the midst of such overabundance, there are fifty million people in America living without access to enough food, in food deserts. At a time when one in six people lacks a secure supply of food to their tables, reducing food losses by just 15 percent would recover enough food to feed more than twenty-five million Americans every year.[58] Wasting perfectly edible food is an ethical issue but also an economic and social problem, with huge implications for the environment.[59] More land does not need to be put under cultivation; rather, more responsible ways are needed for cultivating the land, handling and processing foods, and storing and using food. Food wasted and sent to landfill also contributes to methane gas formation that exacerbates climate change. Improving consumers' tolerance for variation in food products can have an enormous effect on reducing food waste. Freshness and local access through farmers' markets and locally sourced groceries can develop customer loyalty to farmers and lead to greater confidence in the quality of produce. The trend toward prepackaged, processed, and standardized foods increases waste in the supply chain, but customer preferences and unrealistic expectations for perfection create greater waste at the point of consumption.

As climate change and development pressures place new challenges on the food supply system, the pathway to sustainability should be guided by these essential principles:

- Preserve and regenerate fertile ground by practicing organic farming, permaculture, and watershed protection.
- Empower local food resources such as urban gardens, suburban farming, and community-supported agriculture.
- Teach self-reliance—seasonal food, storage and preservation, community care.
- Adopt technology adaptations that conserve resources—greenhouses to extend growing seasons; food waste-to-fuel-to-food systems; cultivation practices that control weeds and insect pests but make minimal use of pesticides, herbicides, and chemical fertilizers.

PITTSBURGH PATHWAYS TO SUSTAINABLE FOOD SUPPLIES FOR A GROWING POPULATION

Pittsburgh is situated in the heart of energy country and has been a hub of fossil fuel development for a hundred years. But Pittsburgh is also in the middle of some of the most fertile land in America. It is a critical foodshed because climate change maps indicate that there will likely be sufficient water to maintain fertile ground here for many years.[60] Here, traditions of farming in many modes, innovation, and strong leadership offer sustainable pathways for food security in a changing world. Many of these initiatives rise from an ethical commitment and a spiritual connection to preserving the life of the earth.

CITY OF PITTSBURGH URBAN AGRICULTURE PROJECT

In Pittsburgh, as in many other American cities and towns, victory gardens were everywhere during World War II, as people patriotically helped to raise their own food so that commercial supplies could go to feeding the army.

When the war was over, the victory gardens slipped into the past, replaced by lawns, commercial foods, and ever more centralized grocery stores. But as illustrated by the productive victory gardens, even in cities there is much capacity for growing food close to where people live. Growing food to help restore the watersheds and to preserve fertile ground could generate a broader connection to the living earth even as it can expand the availability of fresh food to more people.

The home vegetable garden is a tradition that can be recaptured. No matter the size or condition of the space, everyone can connect to the living earth by doing a few simple things. Individuals, families, and communities can:

- Grow lettuce, herbs, tomatoes, or other vegetables in patio pots
- Plant fruit trees as part of the landscape
- Eliminate pesticides and herbicides from the garden and landscape
- Plant seasonal flowers that support pollinators, birds, and wildlife

Rethinking expectations of what grows in residential backyards can include some food as part of the landscape, instead of just grass. When people grow some of their own food, they gain an appreciation of the importance of fertile ground and water, and they understand the food supply chain better. Growing food entails hard work, but the gratification of the harvest brings a sense of satisfaction beyond mere physical nourishment. Recreating a culture of respect and appreciation for fresh food can encourage people to support organic and local choices.

There is no question about the enthusiasm of the city of Pittsburgh for local food and urban agriculture. The deliberate reintroduction of farming within the city has emerged as a way to address environmental justice issues of food deserts, bring healthier and fresher food into the city, and to reclaim abandoned lands. The urban agriculture initiative complements the stormwater management system, which is under review for combined stormwater and sewer runoff. Collecting rain in rain barrels, planting vegetation to help sop up stormwater, and even green roofs and permeable paving emerge as green alternatives to pipes and cisterns.

Shelly Danko-Day serves the City of Pittsburgh in the newly established open space specialist position in the planning department. She has a special interest in developing urban agriculture and making it easier for people to incorporate food production into the urban landscape. In particular, she has been conducting an inventory of vacant and abandoned lots to evaluate the potential for using urban agriculture as a way to restore such places to useful community building functions.

Danko-Day developed a new agriculture ordinance to make it easier for people to have community-supported agriculture operations within the city on lots of two acres or smaller, as opposed to the formerly required three-acre plot size for urban agriculture.[61] If there is a residence on the property, people can keep limited numbers of chickens, beehives, or miniature goats. There are provisions for commercial production of food for wholesale or retail sale. "We are trying to make it easier for people to become more resilient in their food supplies. Community gardens also bring people together and increase the sense of neighborhood," Danko-Day said.[62] To help people develop the skills they need to be successful urban gardeners, the city has partnered with three organizations in town: Grow Pittsburgh, Burgh Bees, and the Penn State Agricultural Extension Service. These organizations provide advice, hands-on assistance, and expertise that is beyond the capacity of the city itself.

Grow Pittsburgh is an urban agriculture nonprofit organization that offers educational programs that teach people at all stages of life and skill about the benefits of growing locally produced, chemical-free delicious, fresh food. "Our mission is to teach people how to grow food and promote the benefits gardens bring to our neighborhoods. We believe access to locally-grown, chemical-free fruits and vegetables is a right, not a privilege. We envision the day when everyone in our city and region grows and eats fresh, local and healthy food."[63]

Grow Pittsburgh started a community gardening program in 2010 to serve as a hub for individuals interested in being a part of this movement, and it has grown quickly. There are more than a dozen thriving community gardens operating in Pittsburgh. One of the most successful is the South Side Slopes. This community garden at Bandi Schaum Field is a collaboration between the City of Pittsburgh Department of Public Works, South Side Community Council, and South Side Slopes Neighborhood Association. On the site of what was once an old baseball field, the garden now boasts eighty-eight plots that can be rented by individual gardeners.[64] "We have a hundred people on the waiting list for a rented plot! So we clearly need more places like this for people to grow food on in the city," says Danko-Day.[65]

Grow Pittsburgh's Edible Schoolyard program operates in schools throughout the city integrating food growing in the schoolyard into science, math, history, and art classes. Carla Garfield, a history teacher at Sewickley Academy, reported on the "Secret Garden" project where the fourth-grade students planted potatoes in three different cultivation methods to see what worked the best. They planted in raised beds with straw, in tubs nestled into the ground, and directly into the tilled soil. In the fall, at the beginning of their fifth-grade year, they harvested fifty pounds of different kinds of potatoes and took them to the cafeteria to be incorporated into lunches served to all the students. One of the students said proudly, "I knew that potatoes grew underground, but I didn't know how they grew or how to harvest them. Now I do!"

In 2012, Grow Pittsburgh launched a formal adult intern and apprentice program to meet rising demand for more hands-on work in urban agriculture across the region. This program is geared toward individuals who aspire to make farming a career in their future. Interested individuals are introduced to the inner workings of a production farm in an urban environment. Training covers all aspects of running the farm, including growing seedlings and produce, maintaining the sites, harvesting food for sale, and selling that product in both a community and restaurant setting. The program offers intern and apprentice positions at Grow Pittsburgh's Braddock Farm, Frick Greenhouse, and Shiloh Farm sites. Since its inception, graduates have taken positions as urban farmers in the Pittsburgh area.

This urban agriculture initiative brings a sense of empowerment to people by connecting the food system to their own community efforts. This process begins to close the gap between people and the natural world. Once the connection to fertile ground and healthy food takes hold, people begin to understand that air and water quality matters, and their own actions can improve the community quality of life. The interconnectedness of living systems manifest in the middle of Pittsburgh neighborhoods formerly filled with blighted buildings brings hope to the whole community.

KRETSCHMANN FAMILY ORGANIC FARMS

The true measure of agriculture is not the sophistication of its equipment, the size of its income or the statistics of its productivity, but the good health of the land.

WENDELL BERRY

Agriculture and food service do not need to be wed to the agroindustrial system to be successful as businesses. The organic food market has grown by double digits since 1996, and local branding based on wholesome, fresh, and sustainably produced food thrives.[66] The Kretschmann Family Farm, on eighty acres near Zelienople in western Pennsylvania, serves two thousand people with organic produce annually. Certified organic through the Ohio Ecological Food and Farm Association and the US Department of Agriculture, Kretschmann Farm was one of the first organic farms in Pennsylvania. Don and Becky Kretschmann were inspired by the words of Louis Bromfield: "Booms must always be paid for one way or another at some time by someone; in the long run there is never any such thing as a 'quick buck.' Someone, perhaps a son or a granddaughter or a child unborn, will have to pay. We are already leaving a vast burden to future generations."[67]

On a warm September afternoon, one field is newly planted with neat rows of radishes and cabbages and kale readying for harvest flourish in another. Colorful rows of flowers, busy with bees and other pollinating insects, form dividers between the fields and border the farm road. Workers travel to the upper fields on a battery-powered adapted golf cart. There, the air is filled by the aroma from a field of heirloom tomatoes bursting with ripeness. These tomatoes will be picked, sorted in the processing barn, and packed into community-supported agriculture (CSA) boxes for delivery throughout western Pennsylvania. Tomatoes produced beyond what can be consumed fresh are processed in small batches into sauce, salsa, and tomato juice for distribution in the CSA boxes, or offered for purchase as extra items.

Since 1985, community-supported agriculture has been growing as a movement. All across the country, farmers have turned to their communities to expand the support needed to have a regular, efficient system of growing organic and local food. The biggest difference between CSA and conventional retail food production, whether at a store or at farmers' markets, is the commitment of the subscriber to a particular farm to purchase a season's worth of produce, rather than an intermittent and unpredictable purchase. That makes a big difference in stability, but also in efficiency. Don Kretschmann said, "Many a time in our years of selling at wholesale or farmers' markets we dumped good produce on the compost pile or didn't pick it because it wasn't or couldn't be sold. Maybe it was a rainy evening at market, or there was a ballgame, or there was a glut at the wholesale level. It's been a sea change to be able to depend on consumers purchasing in this regular way. Since there's a steady group, we really do become a community."[68]

Facing the hillside where the solar power system for the farm is located stands a permaculture orchard of apples grown from root stock and grafted with varieties of apples that resist fire blight and scab. These two diseases account for most of the apple orchard spraying in the East. Because of the careful selection of apple varieties grown to resist disease, the orchard here does not require spraying, and the ground below the trees thrives without constant tilling. Kretschmann Farms provides apples to the CSA shares and produces pressed cider in the fall. Orchards also allow the other forms of soil life free range to colonize, attain balance, and maximum site-specific genetic evolution.[69]

"We have been feeding people healthy organic food since 1971," Kretschmann says with pride. He scoops up a handful of rich ground, noting that his organic practice restores and replenishes the fertility of the ground, so little extra fertilizer is required. He uses a crop rotation process that helps control pests, as well as preserves the soil. Kretschmann scans his land and says: "Preserving fertile ground is an ethical choice for everyone today. We need to reorganize how we do everything so we can live within the resource constraints of the Earth. We hope we are planning adequately for the future and making the sacrifices that might require. Cooperation in loving the children will be the key."[70]

As Kretschmann stood on the edge of the tomato field, he talked about the fate of the farm across the Teats Road border. Across the road from the Kretschmann farm, the land has been sold to a company that plans to install a gas compression station with four to eight compressors on this site. The country road approaching this agricultural area would become an industrial site access road. Kretschmann's legal action to request zoning relief was denied, as was the appeal in civil court.[71] Kretschmann remarks that it is a difficult choice to maintain a farming life under the pressure of the gas boom. Many farmers have leased land or sold out and moved away. He said, "I will be here, and hope my children will carry on the farm for many generations. Our customers depend on us to provide them wholesome fresh food. We are part of this land, and we are here to stay."[72]

EAT'N PARK/PARKHURST: "BUY FRESH, BUY LOCAL"[73]

Started as a single carhop-style restaurant in Pittsburgh in 1949, Eat'n Park Hospitality Group has grown into a portfolio of regional food service businesses focused on personalized dining. Eat'n Park and the Parkhurst Hospitality Group serve more than fifty million customers every year in Eat'n Park restaurants, on college and corporate campuses, and in every state through an online store.[74] Eat'n Park began its sustainability initiative in 2000. The company made significant steps through a corporate commitment to move toward more sustainable business practices, including a suite of actions under the company's EcoSteps program.[75] This sustainability program includes initiatives such as purchasing rBST-free[76] dairy products; recycling spent fryer oil into biodiesel fuel; installing more energy-efficient lightbulbs; switching to cloth rags rather than disposable rags to clean tables; eliminating placemats from all of the restaurants; and replacing paper towels with efficient air dryers in restrooms.[77]

In 2007, Nick Camody, then chief operating officer of Parkhurst Dining Services, accepted the Rachel Carson Legacy Challenge that was part of the centenary celebration of her birth.[78] The company met this challenge with flying colors. Marketing material for the corporation's participation in the challenge notes: "Parkhurst Dining Services, a division of Eat'n Park Hospitality Group, committed to purchase a minimum of 20% of its $36 million in food purchases from farmers within 100 to 150 miles of the communities we serve by December 2008. It converted all of its 'On the Go' containers to corn-based packaging by December 2007. Lastly, Parkhurst promotes Community Supported Agriculture (CSA), small local farmers, by establishing firm corporate and university markets for locally grown products."[79]

Since 2000, Jamie Moore, director of sourcing and sustainability at Eat'n Park Hospitality Group, has been working to expand the company's FarmSource program, an initiative to find and partner with local growers and producers of food. On the business side, the initial driving force behind FarmSource was demand. College students, many with increasing interest in vegetarian diets as an ethical choice, requested local food in their dining halls and Moore helped to develop FarmSource to meet the demand. Moore actively works with about one hundred farms encompassing produce, meat, dairy, and artisanal products sourced within 125 miles of its major institutional customers.[80] In 2014, Eat'n Park Hospitality Group spent about $23 million on locally sourced foods which were served to customers in restaurants, on college and corporate campuses, and in retirement communities and hospitals.[81]

Moore believes education and relationship building are critical to the success of the FarmSource program. For instance, in 2012, a group of Eat'n Park Hospitality Group servers and chefs spent a day at a local farm. The group got their hands dirty and planted seeds for six hundred pepper plants, transplanted six thousand zucchini plants, and picked blueberries. Employees saw where the food grows, and they could share that experience with their customers. Moore finds that this "transparency with customers creates trust—a sense of knowing where their food is coming from."[82] He continues to build trust by visiting growers and collaborating to add local foods to the menu.

Eventually, Moore would like to create a local food section on the Eat'n Park Restaurant menu. This would require teaming with growers to understand what that they can grow in large quantities "that will work with the menu and bring new variety to it." With more than seventy Eat'n Park restaurant locations throughout Pennsylvania, West Virginia, and Ohio,[83] adding a new locally sourced menu item is a big job; yet organizing new initiatives, visiting farms, and getting things moving is something that Moore enjoys about his work.

In addition to cultivating demand for local foods, developing an aggregation point for local growers to bring local products to market is critical to successful local sourcing. "Chefs live or die by food costs and you need a transportation system to get reasonably priced, high-quality local food into the market."[84] Regional aggregation points keep transportation costs down and reduce produce damage from shipping over longer distances. Products reach their destination faster and with less handling.

An ethic of preserving the fertility of the land is emerging as part of corporate social responsibility and sustainability objectives. Kevin O'Connell, senior vice-president of marketing, believes that "if you have the right goals, you'll find new ways to bring fresh food to the forefront. In 2009, we decided we wanted to be a restaurant leader in sustainability and pursue a LEED Gold certification for our newest Eat'n Park Restaurant."[85] The restaurant, located in the Fox Chapel neighborhood of Pittsburgh, incorporates many unique elements, including: ENERGY STAR–rated kitchen appliances, rainwater barrels to water the restaurant landscaping, specially designed skylights to amplify natural light, twelve-watt LED lighting through the restaurant, and environmentally friendly and recycled construction materials. The most visible sustainable element is the forty-foot wind turbine attached to the roof. This is Pennsylvania's first wind turbine system for a restaurant, and it generates two thousand kilowatt-hours of clean energy per year. "The next time you are in," Moore said, "make sure to look for our 'touch points,' which highlight the restaurant's sustainable features."[86] As many people come to the restaurant and see the energy-efficient features and information about how they are integrated into the operation, they gain awareness of sustainability in the food system.

CONCLUSION

Agriculture is caught in a time of necessary shift away from a system based on fossil-fueled machines, petroleum-derived chemicals, and monoculture on a grand scale. The way forward progresses slowly, the way an abandoned airfield moves from paved asphalt to a meadow, then to a shrubby field, and finally to a forested land. The new farming movement grows through the person-to-person reconnection with the land. It grows through people wanting more food that is healthy and fresh. It grows through people taking back abandoned city lots and making food flourish in formerly blighted plots.

People can live within the ecosystems they inhabit without destroying the essential regenerative balance of the cycle of growth, decay, and regeneration. Farming based on the principles of nature rather than on the pesticide treadmill and the endless escalation of chemical-dependent plants and animals does require more labor. Not the impersonal semislavery of low-wage workers migrating from plot to plot in seasonal drudgery, but the labor of people tied to the land with community and cooperation for the benefit of all and the use of each.

No technological revolution is necessary. Urban areas can flourish with food, trees, and flowers rather than barren concrete enclosures for harried people. Rehumanizing the food system to grow food for people, rather than mostly for animals and commodity markets, will restore fertile ground. Educating people, especially Americans, about the ethical implications of their food choices can greatly improve the capacity to feed people everywhere. If people have an interest in their own sustenance, they will recognize that the living earth will reward the effort of its cultivation.

Lalit Cordia, entreprenuer, founder, and CEO of Thar Industries, focuses on green energy solutions.

THAR INDUSTRIES

6
PREVENTING POLLUTION

We have subjected enormous numbers of people to contact with these poisons, without their consent and often without their knowledge. . . . We have allowed these chemicals to be used with little or no advance investigation of their effect on soil, water, wildlife and man himself. Future generations are unlikely to condone our lack of prudent concern for the integrity of the natural world that supports all life.

RACHEL CARSON

IN THE SPRING of 2008 researchers from the University of Pittsburgh Graduate School of Public Health stood on a dock on the Allegheny River scanning the current for the best place to catch fish. With them were local high school students, enthusiastic to begin collecting fish for the Allegheny River Stewardship Project.[1] As part of the participant observer research team, students caught fish, recorded data on their specimens, and preserved them for further analysis in the laboratory. The research using fish as biomarkers to study the effects of heavy metals included the local knowledge and experience of the Allegheny River shared by the student participant researchers. Two high school boys tugged a graduate researcher by the sleeve and said, "You won't catch any fish here. They are down a bit at that curve, but you can't get there from this dock."[2] The boys led the way through the woods and down the bank where the water formed a small eddy and a calm pool beneath overhanging trees, their own special place for fishing. They each caught several fish, filled in their data sheets, and headed back to tally up. They reported that they regularly catch fish there for the family to eat.

All summer, researchers and students went to places along the river that locals identified as good fishing areas. Residents talked about problems they noticed, such as white material leaching out of the stream bank into the water, or places where the sewer outfalls emptied. When all the data were compiled, the researchers discovered that in some places, over half of the fish samples showed signs of endocrine disruption, tumors, or other physical malformations.[3] The most frequent tallies of these malformations were found in areas where sewer outfalls flowed into the river. On analysis, fish showed traces of toxic materials like PCB—which was banned in 1972—and levels of mercury 25 percent higher than levels the EPA considered safe for human consumption.[4] What is going on here? "Whatever you throw down your drain or into your toilet goes into the rivers," explained study director, Conrad Daniel Volz.[5]

Manmade chemicals have become part of everyday life, and part of every living organism, including people. Synthetic materials define the modern lifestyle. Convenience is expected in everything from dishes, tableware, and medical supplies to manufactured products and packaging. Mass production with inexpensive components derived from petroleum, coal, and natural gas has propelled the consumer economy with "better living through chemistry." Increased use of synthetic materials has had the unintended consequences of negative health effects from contamination of the environment and from direct exposure to humans. People see media advertisements for new products to kill "weeds" or "pests" or perform other convenient functions, but there are no clear discussions of the harm or health effects of such products. The product labels on foods, personal care products, cleaning products, and pharmaceuticals offer cautions of "side effects," but busy consumers usually ignore such warnings, especially if the specific harms associated with particular chemicals are not well known.

The Centers for Disease Control and Prevention regularly monitor human exposure to synthetic chemicals through a biomonitoring study sampling blood and urine from a statistically valid number of the adult population nationwide. The most recent study shows that average Americans have 441 synthetic chemicals in their bodies, 79 of which are known to cause cancers, mutations, or developmental disorders.[6] One such chemical, Bisphenol A (BPA), was found in 93 percent of persons tested.[7] BPA is found in a wide variety of products, including linings of cans for food, food packaging materials, beverage containers, and items such as compact discs, plastic dinnerware, and high-impact protective gear such as helmets and some car components.[8]

People are exposed to BPA from eating food or drinking beverages stored in BPA containers; children can be exposed from toys or drinking containers that they put in their mouth. Inhalation exposure can also occur, especially for manufacturing workers. A study conducted by the National Institute of Occupational Safety and Health found levels of BPA in the urine of manufac-

TABLE 6.1. MOST COMMON ENDOCRINE DISRUPTORS

AGRICULTURE / FOOD
Bis-Phenyl-A Pesticides / herbicides: Heptachlor, atrazine, organophosphates
ENERGY PRODUCTION
Dioxin Heavy metals: Cadmium, arsenic, mercury, lead
CONSUMER PRODUCTS
Pthalates Parabens Polybrominated diphenyl ethers (PBDE) Polychlorinated biphenyls (PCB)

SOURCE: DR. LINDA BIRNBAUM, NATIONAL INSTITUTE FOR ENVIRONMENTAL HEALTH SCIENCES, 2009

turing workers to be one thousand times higher than those found in the general population.[9] Their exposure pathway included both inhalation and contact. Another occupational exposure to BPA comes from workers who handle thermal paper receipts that are often coated with BPA. An Environmental Working Group study found that retail workers had 30% higher levels of BPA in their bodies than the general population.[10] Chemicals that cause endocrine disruption, neurologic disorders, and immune system disorders as well as cancers are everywhere in materials people use every day. Considerable evidence suggests that some of the maladies that appear later in life may be consequent to early exposure to endocrine disruptors. Table 6.1 illustrates the most common sources for endocrine disruptors.

Rachel Carson's warnings from 1962 ring true today: "We are accustomed to look for the gross and immediate effect and to ignore all else. Unless this appears in such an obvious form that it cannot be ignored, we deny the existence of hazard."[11] The cumulative evidence that health is affected by contaminants is hard to ignore:

- The United States has one of the highest cancer rates in the world.[12]
- One in six children has a developmental disorder.
- Childhood asthma more than doubled in twenty years, and now affects one in four Americans.
- One in every eight babies is born prematurely.
- Sperm count is falling at 1 percent a year.[13]

It is difficult to attribute these effects to any single pollutant or to any specific combination of pollutants. The issue of health effects from environmental contaminants is further complicated because not every person responds to the same chemical in the same way. The response to exposure from substances in the environment depends on genetic predisposition and the timing of the exposure. Some exposures may be harmless to adults but could be devastating to a developing child, and especially to the fetus. In some cases, exposures early in life are not manifest until later, as with some cancers that may take years to develop. Other effects may not show up until the next generation.

In spite of regulations limiting the amount of toxic or hazardous material that can be released into the environment, millions of pounds of chemicals are released annually. And unlike pharmaceuticals and pesticides, which must be subject to extensive health studies to determine their safety before they can be commercially marketed, industrial chemicals are considered safe until they are proven to be hazardous. Furthermore, the Toxic Substances Control Act of 1976 (TSCA) that was introduced as a regulatory constraint "grandfathered in" approximately sixty-three thousand chemicals already in commercial use at that time, and applied restrictions to five, with an additional four categories of chemicals limited in more recent times. As a result, most Americans have accumulated hundreds of synthetic chemicals in their bodies, many known to cause mutations or cancers. Even newborn infants show this pattern of bioaccumulated chemical contaminants.

Although some of the thousands of new compounds added to commercial production annually may be inherently innocuous, there are many whose biological interaction is potentially harmful. It is more alarming that the standards for evaluating health effects and public safety are set in isolation with specific "safe levels" determined for each compound of concern. Unfortunately, living systems do not encounter one chemical at a time but rather are bathed in a mixture of chemicals of variable and unknown effect. The chemical stew that surrounds every person comes from contaminants in the air, water, food, personal care products, home care products, clothing, and even furniture. Exposure is ubiquitous and unavoidable.

COMMERCIAL CONTAMINANTS

> *For the first time in the history of the world, every human being is now subjected to contact with dangerous chemicals, from the moment of conception until death.*
>
> **RACHEL CARSON**

In *Silent Spring*, the seminal work on the topic, Rachel Carson characterized the problem of chemical contamination: "There is still very limited awareness of the nature of the threat. This is an era of specialists, each of whom sees his own problem and is unaware of or intolerant of the larger frame into which it fits. It is also an era dominated by industry, in which the right to make a dollar at whatever cost is seldom challenged."[14] Thus the broad use of pesticides and herbicides continues today, more than half a century after Carson's warning, even in the face of evidence that critical segments of the ecosystem, such as pollinators and butterflies, are doomed by the practice.

Today, within the allowances of existing law, 5.2 billion tons of toxic chemicals are released into the air, water, and land each year in the United States alone.[15] Of approximately eighty thousand synthetic chemicals in common commercial production today, fewer than two hundred have been tested for health effects, and only nine have been banned or limited as posing an "unreasonable hazard" under the TSCA.[16]

Americans are the largest consumers of goods in the world, including personal care products, manufactured goods, pharmaceuticals, household and industrial chemicals, and industrial and domestic agricultural chemicals. Many products banned for use in Europe due to safety concerns are manufactured there, or in countries with lower standards for quality and contaminants such as China, and sold exclusively in America. One such compound is atrazine, banned for use in Europe from concern about evidence that it interferes with reproduction and development, and may cause cancer.[17] But millions of pounds of atrazine, manufactured by Syngenta, a Swiss company, are applied to cornfields each spring all across America. Similarly, the World Health Organization has determined that the herbicide glyphosphate, manufactured and marketed by Monsanto under the name Roundup, is a "probable carcinogen" in humans.[18] Roundup is the most heavily used herbicide in the United States, with 180 to 185 million pounds applied annually to commercial agriculture and five to eight million pounds annually used in domestic applications, but the EPA still has not yet fully evaluated it for risk to humans.[19] Millions of people are exposed to these materials, often unknowingly or unwillingly. The effects on wildlife such as insect pollinators, amphibians, and soil fungi are discounted in regulatory evaluations, but have potentially catastrophic consequences for ecosystems.[20]

Convenience also pervades the market for treating surfaces to give them water-repellent or other properties. The Madrid Statement, authored by fourteen international researchers and signed by two hundred scientists, calls for regulation and replacement of perfluoridated compounds widely used in waterproofing, stain-resistance coatings, and nonstick applications since the 1960s. These materials degrade very slowly, if at all, under ambient environmental conditions. They are found in the indoor and outdoor environments, wildlife, and human tissue and bodily fluids all over the globe. They are emitted via industrial processes, military, and firefighting operations, and migrate out of consumer products into air, household dust, food, soil, ground and surface water, and make their way into drinking water.[21] "Stain-resistant, nonstick, waterproof, and lethal" is how one writer referred to a highly fluorinated chemical found in the bodies of nearly all humans and animals on the planet.[22] This material, C8, is the primary component of nonstick cooking pots and pans, and fabric treatments and surfaces. Exposure to it is linked to cancer, liver malfunction, hypothyroidism, obesity, decreased immune response to vaccines in children, hypertension, and other health problems.[23]

In March 2008, the Associated Press reported that pharmaceutical compounds were detected in the drinking water of twenty-four major metropolitan areas across the country serving forty-one million people.[24] This information was derived from tests that water utilities had undertaken voluntarily and provided to the press. Detected drugs—residues of medicines taken by people that end up in the waste stream—included antibiotics, anticonvulsants, and mood stabilizer drugs. These results supported previous findings by the US Geological Survey, which sampled 139 streams in thirty states in 1999–2000 and found organic wastewater contaminants and pharmaceuticals in 80 percent of sampled sites—including antibiotics, hypertensive and cholesterol-lowering drugs, antidepressants, analgesics, steroids, caffeine, and reproductive hormones.[25]

REGULATIONS IN THE UNITED STATES

As more information emerges documenting the health effects of some chemicals with long-lived stability in the environment, the effects on future generations and nonhuman life become important considerations. The unrestrained production of chemicals, some with long-term biological toxicity, affects entire ecosystems.[26] Understanding the critical interconnectedness of living systems reveals the devastating effects manmade materials can have on viability. In an ethical society, the laws would protect the unborn, the economically disadvantaged, and nonhuman living things above the profits of individual corporations. How can we change the laws governing the chemical production system that have been entrenched for over fifty years? Balanced risk has become the norm. But balanced risk does not give voice to the unborn, or to the nonhuman species whose lives and viability are affected.

Existing laws and regulations in the United States make it difficult to enforce prohibitions of chemicals without specific proof that a particular chemical causes a specific harm to a significant number of people. Regulatory agencies have only the powers and jurisdiction granted by the laws that created them, and they require due process before rules or restrictions can be adopted. When regulators seek comments and input from the public, the most detailed and intensely represented positions are those of the affected industries. Laws often explicitly require balancing of economic, environmental, cultural, and historical interests in defining the parameters for a decision. Regulators are required in the enabling legislation of TSCA to consider the economic impact on the manufacturers when they consider the need to ban or replace specific chemicals.

One example of the consequences of this system of regulation is the persistence of asbestos. Even though the health effects of asbestos are well known, and exposure has been linked to causing a form of cancer called mesothelioma, the substance has not been banned. In 1989, the EPA issued a final rule under Section 6 of TSCA banning most asbestos-containing products. However, in 1991, this rule was vacated and remanded by the Fifth Circuit Court of Appeals because the company had no economic alternative, and the law was not empowered to put a company out of business. As a result, most of the original ban on the manufacture, importation, processing, or distribution in commerce for the majority of the asbestos-containing products originally covered in the 1989 final rule was overturned.[27]

After this definitive ruling of the court, it became clear that the enabling legislation must be changed to allow more stringent enforcement under TSCA. When a clearly harmful product remains commercially available because of limitations in the law, the action may be legal, but it is not ethical. Attempts at comprehensive TSCA reforms have been foiled in Congress for many years, but the Frank R. Lautenberg Chemical Safety for the 21st Century Act was finally signed into law on June 22, 2016.[28] It emphasizes voluntary compliance approaches, with industry self-reporting and self-regulating production as the centerpiece. The law implements several essential reforms: mandatory requirement for the EPA to evaluate existing chemicals with clear and enforceable deadlines; new risk-based safety standard; increased public transparency for chemical information; and a consistent source of funding for EPA to carry out the responsibilities under the new law.[29]

The public health standard, "a reasonable certainty of no harm," applied to foods, pharmaceuticals, and pesticides, ensures that a chemical cannot pose more than a one in a hundred thousand to one in a million risk of developing cancer over a lifetime of exposure. Unfortunately, current legislative initiatives for commercial chemicals are based on the more permissive industry standard.[30] Based on "no unreasonable expectation of harm," it allows latitude for industry to assume that some harm can be acceptable within economic parameters, and is less protective of the consumer. In the absence of industry mandates to take precaution with public health and environmental protection, the responsibility falls primarily on the consumer to avoid buying products that have chemicals of concern.

Poor labeling and a general lack of knowledge about harmful effects complicate the challenge. Manufacturers control the power of the market and argue that people should be allowed to make their own choices. But if consumers do not know what chemicals are in food products, personal care products, or household cleaners, how can they make better choices? In some instances, industries have suppressed detailed labeling to prevent consumers from avoiding their products.[31] In other instances, companies may make a "green" product that has blatantly not reformulated, but may be marketed differently to attract customers. Products with "natural" or "organic" labels are not required to eliminate or specify materials such as fragrances, inert ingredients, coloring, or preservatives. Toxic materials used in packaging, or liners to canned goods are not indicated on the label either. Even if such labeling were required, however, labels with long lists of chemicals in very small print are not really informative to a consumer unless the consumer knows which compounds have health effects. Controlling the information available to consumers limits their informed freedom of choice.

For the American consumer, there is no way to avoid exposure to hundreds of synthetic chemicals. Even diligent care in shopping does not assure safety, because many materials are not included in the labels, and many are inhaled or ingested through air and water pollution that cannot be avoided. Banning a larger and larger array of chemicals as "proof" of harm becomes documented will not solve the problem. Even enforcing the existing laws more stringently only reduces some of the difficulty. It is clear that a different way forward is necessary to begin remediating this problem. An entirely different approach is needed to limit risks by designing products and processes to avoid the hazard from chemicals that cause toxicity, cancer, mutations, endocrine disorders, or developmental disorders, rather than limiting the exposure through permitted emissions. This arena of health effects from chemical production, use, and disposal most urgently needs public policy actions on behalf of the all exposed living things as well as the human public interest.

A PREMISE OF PRECAUTION

Most environmental regulations adopted between 1969 and 1990 focused on specific limits on what could be emitted in the air, discharged into water, or buried in landfills. The assumption of the time was that "the solution to pollution is dilution," based on the long-held principle of toxicology that at some point of dilution, there will be no detectable response to a toxicant. The regulatory premise founded on the assumption that the dose makes the poison traces back to the experiments of Paracelsus, considered the founder of toxicology, in the 1400s.[32] People have assumed that the ocean was so deep, the sky so vast that no human activity would have a lasting effect on the overall environment. Between 1970 and 1976, the environmental protections of the Clean Air Act, Safe Drinking Water Act, Resource Conservation and Recovery Act, and Toxic Substances Control Act set a foundation for environmental protection by putting regulatory corks in smokestacks, stoppers in emission pipes, and liners in landfills, but they did not look at the pollution-causing systems to seek out long-term solutions.

The decade from 1970 to 1980 captured a rare period of general consensus about government policy to protect the environment and the public interest. By the middle of the 1980s, however, the regulatory shift in favor of free markets took hold with strong support from industries. People fear that if industries are regulated for emissions to the environment, their own freedom to use things like pharmaceuticals, lawn pesticides, and fertilizers will be curtailed as well.[33] The environment has become a battlefield for competing interests

TABLE 6.2. ENVIRONMENTAL PROTECTION IN US LAW, 1970–1990

1969 National Environmental Policy Act
1970 Clean Air Act
1970 Occupational Safety and Health Act
1972 Coastal Zone Management Act
1972 Marine Mammal Protection Act
1972 Clean Water Act
1972 Noise Control Act
1973 Endangered Species Act
1974 Safe Drinking Water Act
1976 Fisheries Conservation Act
1976 Toxic Substances Control Act
1976 Resource Conservation and Recovery Act
1980 Comprehensive Environmental Response, Compensation, and Liability Act
1982 Nuclear Waste Policy Act
1986 Emergency Planning and Community Right-to-Know Act
1988 Marine Protection, Research, and Sanctuaries Act
1990 Oil Pollution Act
1990 Pollution Prevention Act
1996 Federal Insecticide, Fungicide, and Rodenticide Act

SOURCE: US ENVIRONMENTAL PROTECTION AGENCY, WWW. EPA.GOV

pitted against each other and against governmental regulations. Court challenges have yielded contradictory rulings, especially as states began to interpret the regulatory impact on a case-by-case basis.

The current system allows the production and sale of chemicals as long as there is no direct "proof" that they are harmful. The insistence on undisputed proof has become the tool of industries to derail effective regulation of the production and use of most industrial processes or products.[34] In the United

States, the industry data is protected as proprietary information, and proof of harm rests with the government, or consumer advocates with a high bar for regulatory action. The majority of regulatory controls address the allowable level of exposure considered to be safe for workers, the public, and wildlife. The regulatory system assumes that if the exposure levels are controlled to low enough amounts, the public will generally be safe from toxic effects. However, as scientists learn more about how natural systems function and study the effects of some chemicals even at very low exposure levels, these legacy regulatory assumptions have come into question.

Many countries use a regulatory approach based on a premise of precaution, in which new chemicals may not be released for general production or use until they are shown to be safe.[35] In Europe, the precautionary principle was adopted as the basis for its chemical regulatory system, called REACH—an acronym for Registration, Evaluation, Authorization and Restriction of Chemicals.[36] In this system, industry has greater responsibility to demonstrate with specified data that chemicals are not harmful to health before they can be marketed, and if a chemical is in a category of very high concern, a substitute material must be identified.[37] This differs from the US regulatory process in two important respects: first, the burden of proof of safety lies with the manufacturer, not the government agency; and second, until safety is demonstrated, chemicals may not be sold to the public, as a precaution against unintended consequences.

The practice of preventing pollution, rather than trying to contain its effects, addresses a critical ethical issue of taking precaution when effects of unknown result affect the unborn as well as nonhuman living things, and in exposing people to substances that may be harmful without their knowledge or consent. Rachel Carson advocated this position, especially regarding the potential for genetic damage. "I don't believe we should wait for some dramatic demonstration before making a thorough study of the potential genetic effects of all chemicals that are widely introduced into the human environment. By the time such a study is made otherwise, it will be too late to eradicate them."[38] The pathway to a sustainable future for materials and products follows the principles of green chemistry, which limits the risk from chemicals by designing products and processes that are inherently nontoxic, and uses biomimicry, following the example of natural systems.

GREEN CHEMISTRY SOLUTIONS

> *Nature manufactures its materials under life-friendly conditions—in water, at room temperature, without harsh chemicals or high pressures. Despite what we would call "limits," nature manages to craft materials of a complexity and functionality that we can only envy.*
>
> **JANINE M. BENYUS**

Chemists enjoy a challenge, especially one that expands the boundaries of their expertise. Usually, chemists and chemical engineers focus on the performance of the product, its longevity, and the economy of the supply chain. Less frequently are they concerned with how the chemicals break down, or whether they or their degradation products are toxic to living systems. A chemist or chemical engineer may not consider whether a chemical product might persist in ways that can accumulate in food chains. When the effect of chemicals on living systems becomes part of the discipline necessary for chemists and chemical engineers, they will have the tools to design products and processes to eliminate the hazard.

As a critical criterion to reduce chemical hazards, chemists need to avoid creating compounds with toxic effects on living systems when they design materials, products, and processes, thus avoiding the production of biologically harmful products or byproducts.[39] To achieve this, chemists' education needs to include an understanding of biology, especially toxicity and ecotoxicity, ecology, physiology, and endocrinology.[40] They need to know what happens to chemicals that enter the ecosystem and understand how chemicals act in the bodies of humans and other living things. These concepts are integral to the practitioners of green chemistry.

Green chemistry is the design of chemical products and processes that reduce or eliminate the use or generation of hazardous substances. Green chemistry applies across the life cycle of a chemical product, including its design, manufacture, use, and ultimate disposal.[41] The concept has its roots in the Pollution Prevention Act of 1990, which established a four-part pollution prevention hierarchy as federal policy:

1. Pollution should be prevented or eliminated at the source, by design.
2. Pollutants should be recycled whenever possible.
3. If materials cannot be recycled, they should be treated in an environmentally safe manner.
4. Any residual material should be disposed in an environmentally safe manner.

The process for implementing these provisions evolved as the discipline of green chemistry. In practice, green chemistry prevents the formation of hazardous chemicals in production processes or the products that reach consumers or the environment. The major weakness of this statute lies in the weak enforcement provisions, and the lack of a mandate for these principles to apply to existing processes. Thus, even though technologies for making less-harmful products exist, they are not in the mainstream of use in producing consumer goods.

Preserving the biosphere as a habitat for humans and all other living things is a critical priority of sustainability. This problem of the unintended conse-

TABLE 6.3. THE PRINCIPLES OF GREEN CHEMISTRY

1. Prevent waste
2. Design safer chemicals and products
3. Design less hazardous chemical synthesis
4. Use safer solvents / reaction conditions
5. Increase energy efficiency
6. Use renewable feedstocks
7. Design chemicals and products to degrade to benign substances after use
8. Minimize the potential for accidents
9. Analyze in real time to prevent pollution
10. Use catalysts not stoichiometric reagents
11. Maximize atom economy
12. Avoid chemical derivatives

SOURCE: US ENVIRONMENTAL PROTECTIONS AGENCY, WW.EPA.GOV

quences of synthetic chemicals daunts the most dedicated advocates. An effective response requires united action across many disciplines and bridges several sectors. Several innovative companies and institutions in Pittsburgh are pursuing this essential discipline, and working to make green chemistry practices a mainstream way of doing business.

Pittsburgh presents a rare convergence of forward-thinking expertise in health, environmental and medical research, academic excellence in green chemistry, multiple industrial research and applications companies, a supportive philanthropic community, and a global network of collaborators linking inspired leaders who have pioneered interdisciplinary scholarship, research collaborations, and educational reform. Local leadership has helped to build this broader collaborative space to focus the collective wisdom of chemistry, biology, endocrine disruption science,[42] and systematic approaches to sustainability on the design of products, processes, and activities that are benign to health and the environment.

TAML CATALYSTS FOR WATER PURIFICATION: BIOMIMICRY IN PROCESSES

The US Geologic Survey's biannual analysis of the nation's drinking water supplies documents the presence of many chemicals, including detergents, pharmaceuticals, plasticizers, emulsifiers, pesticides, and petroleum products.[43] In recent sampling, radioactive isotopes and high levels of dissolved salts extracted with the recovered fluids from hydraulic fracturing are also turning up in drinking water supplies.[44] These materials are not removed by filtration through sand filters or even carbon filters. The typical water disinfectant method of using chlorine to kill bacteria, followed by aerial oxygenation or ultraviolet light purification, may allow some volatile hydrocarbon materials to escape, but heavier or more complex compounds are not removed. Nor are materials that completely dissolve in water, such as salts, affected. "It doesn't matter if it's a large urban wastewater plant, a midsize city wastewater plant or individual septic tanks. These chemicals are present," according to Larry Barber, a geologist with the USGS.[45]

Protecting public water supplies faces the challenge of how to remove increasing amounts of complex organic materials. Contamination with endocrine-disrupting chemicals poses special concerns. The two currently dominant technologies for public water supply treatment use ozone and activated carbon, but there is real promise for new approaches based upon small molecule catalysts that mimic oxidizing enzymes found in nature.

Terrence Collins, Teresa Heinz Professor of Green Chemistry and the director of the Carnegie Mellon University Institute for Green Science in Pittsburgh, has been working on the problem of removing organic contaminants such as endocrine disruptors from the water supply. Collins says, "Endocrine-disrupting chemicals change the meaning of chemicals to the human journey. We now understand that some everyday/everywhere chemicals can disrupt hormonal developmental and physiological signaling processes to impair organisms with endocrine systems, including humans. The disruptions can occur at environmentally relevant low doses and are marked with the complexities of time-of-exposure dependence, nonmonotonic response behavior."[46] This is a particularly important problem because the usual processes for treating municipal water supplies are designed to remove debris by filtration and destroy bacteria, usually with chlorine. Neither of these procedures will remove dissolved chemicals or complex organic compounds.

The natural systems for managing toxins have coevolved with our habitat over thousands of years, so in seeking answers we can look to how natural systems handle toxic materials. Collins says, "I always look to nature for inspiration." He explains that natural processes draw the building blocks for biosynthesis from naturally occurring and abundant components that can combine through quite elegant and complex reactions in living things. Think of the crystalline structure of corals, or the variety and complexity of insect pheromones for communication among members of an ant colony, for example. Natural systems also break down the products that are formed so that they return to basic elements or very simple molecules, such as carbon dioxide and water for reuse and redistribution. In natural systems, such as in plants and animals, enzymes are the heavy lifters for biosynthesis and biodegradation. All of nature's systems run in cycles and feedback loops that support specific functions, then regenerate the raw materials.

Modern synthetic chemistry does not operate in this biological way. Modern chemical industries commonly begin with feedstocks from fossil petroleum, coal, or methane and build these into bigger molecules and complex polymers and compounds for which there are no natural analogues, and natural enzymatic systems are often unable to effectively degrade these back to the simple molecules of the resource base. Collins notes, "Some of the new synthetic molecules in medicines, plastics, and pesticides are so different from the products of natural chemistry that it is as though they dropped in from an alien world. Many of these molecules do not degrade easily, and even some biodegradable compounds have become omnipresent because we use them so copiously."[47]

Collins picked up on scientific findings in the 1980s that the chlorine used in municipal water treatment was combining with organic chemical contaminants to create toxic byproducts of the chlorinated disinfection process. This concern, combined with his growing awareness of the endocrine-disruptive properties of many synthetic chemicals and their byproducts, launched his search for a better system to remove contaminants from drinking water. "Rather than relying on chlorine, we wondered if we could put nature's own cleansing agents—hydrogen peroxide and oxygen—to the work of purifying water and reducing industrial waste. These cleansers can safely and powerfully obliterate many pollutants, but in nature the process usually requires an enzyme—a biochemical catalyst that vastly increases the rate of the reaction," Collins noted.[48] But massive amounts of enzyme would be needed in a water purification application. Finding a way to emulate the elegant and complex functions of natural enzymes proved to be a daunting task.

What emerged as a solution after fifteen years of trial and error was a series of designer catalysts built around an iron atom that serves as the catalytic site that would work to break down the organic chemical contaminants. These designer catalyst molecules, called TAML activators, work with hydrogen peroxide and other oxidants to break down a wide variety of stubborn micropollutants. "TAML catalysts accomplish this task by mimicking the enzymes in our bodies that have evolved over time to break down chemicals by oxidizing them," Collins says. "In laboratory and real-world trials, TAMLs have proved they can destroy dangerous pesticides, dyes and other contaminants, greatly decrease the smells and color from the wastewater discharged by paper mills, and kill bacterial spores similar to those of the deadly anthrax strain. If broadly adopted, TAMLs could save millions of dollars in cleanup costs."[49]

TAML catalysts are the first highly successful synthetic replicas of peroxidase enzymes. Such enzymes in living things activate hydrogen peroxide to oxidize organic compounds in ways that are reminiscent of combustion. A similar reaction occurs when TAML catalysts combine with hydrogen peroxide to produce powerfully oxidizing intermediates that readily degrade oxidizable chemicals. The active agent faithfully mimics the functions of enormous protein enzymes but with small molecules based on iron attached to four nitrogen atoms held together by a molecular circle of organic composition that activates the peroxide to oxidize the atoms and the bonds of the contaminant target, effectively producing "fire in water."[50] The TAML catalysts remain intact and functional for several hours, long enough to run around the catalytic cycle large numbers of times, but not to the point of being indestructible so they would not become contaminants or pollutants themselves.[51]

A TAML-enabled oxidation rapidly kills hardy pathogens in water, including anthraxlike spores, the hardiest of all. Problematic chemicals degraded by TAML-driven oxidation include endocrine disruptors (EDCs), chlorinated pollutants, phenols, dyes, and chemical warfare agents. EDCs destroyed include the active ingredient in the birth control pill—ethinylestradiol.[52] Industrial targets include thiophosphate and organochlorine pesticides, the odorous and colored compounds in pulp mill effluent, and—among other things—the recalcitrant sulfur compounds in diesel and gasoline. And TAML plus peroxide has excellent potential for dye transfer inhibition and stain removal in laundry products and for increasing the whiteness, brightness, and freshness of the washed fabrics.[53] Currently the TAML catalysts are in use in several industrial applications. "The power of this catalyst is that you need very small quantities to treat large volumes in production," said Colin Horowitz, the chief technology officer for GreenOx Catalysts, the company that has been marketing a TAML catalyst.[54]

In the Institute for Green Science at Carnegie Mellon University, Collins's research group continues testing and refining the TAML catalysts with the goal of using this process for municipal water purification.

> Endocrine disruptors present a powerful threat to the welfare of all living things. We are engaged in two strategies to minimize the damage of endocrine disruption to the future. In the first of these, a team of environmental health scientists has produced the Tiered Protocol for Endocrine Disruption, or TiPED, which collects the most powerful assays for identifying and understanding endocrine disruption behavior in chemicals. The TiPED is a power-

> ful tool for steering the design of new chemicals away from endocrine disruption. In the second, we are showing that TAML activators can underpin high technical, cost, health, and environmentally performing processes for removing endocrine disruptors that society is using from water.[55]

Collins's work springs from his deep concern for future generations and for reshaping the impact that chemicals and materials have on the environment and the living things to which they are exposed. As a precaution against long-term harms that accumulate in living systems, chemicals must be designed to do no harm through the long life cycle of each chemical from its creation to its final decomposition to elemental substances.

BIOMIMICRY IN MATERIALS

New products drive the market with innovations for increased performance or improved customer convenience. Technology applied to create bigger, better, faster solutions constantly churns the production system. Often, the innovations so produced complicate the environmental impact with byproducts that have toxic or hazardous properties. However, some solutions offer improved performance and convenience without increasing the environmental burden. Using biomimicry as inspiration opens many possibilities for improved performance without increased contamination. Two Pittsburgh companies are presented here based on their ethical decisions to avoid pollution as part of their corporate philosophy.

COHERA MEDICAL, INC. A BIOMIMICRY SUCCESS STORY

Eric Beckman is a polymer chemist who looks at the world from a totally pragmatic perspective. For instance, at the Rachel Carson Green Chemistry Roundtable in 2009, addressing chemical contaminants, Beckman was asked about shifting from a gasoline-powered lawn mower to an electric one as a way to reduce air pollution. Beckman's answer framed the issue in a new perspective when he replied: "Innovation can make leaps to totally change the approach in ways that cannot be achieved with incremental change. We need to innovate by subtraction. For example, take oven cleaner. There is no way to 'improve' oven cleaner to be less toxic, but the self-cleaning oven does the job!

This is a total innovation to solve the problem. Another example: people work on having a greener lawn mower, but you can plant low-growing grass instead."[56]

Many synthetic products rely on petrochemical source materials and solvents and are developed through processes that generate toxic air and water releases. Disposing and treating wastes represents a cost to the industry both in the form of wasted material and additional procedures to clean or detoxify emissions to the extent required by regulations. Even with controls, air and water pollution results from the cumulative effect of permitted releases.[57] Beckman prefers an approach that looks at meeting a customer's need for performance while preventing pollution by how the production process is designed. "You've paid money for materials, you've paid money to process the materials, and now you are going to pay money to throw part of them away? That doesn't make any sense economically."[58] He has found inspiration in natural systems, especially high-performance natural polymers: "Where in nature do you find strong adhesives that do not dissolve in salt water? Anyone who has ever had to scrape barnacles from the hull of a boat knows the strength of adhesive that holds marine barnacles to the rocks, or adheres them to a surface."[59]

This capacity to ask questions in a different framework, seeking solutions to the underlying problem rather than the addressing the symptoms, characterizes Beckman's work. Beckman received a Green Chemistry Challenge Award from the EPA in 2002 for his work on designing materials that could be soluble in liquid carbon dioxide, a synthesis design breakthrough. He founded and remains as codirector for the University of Pittsburgh Mascaro Center for Sustainable Innovation to cultivate leadership and inspire young people to develop "innovations that benefit the environment, positively impact the community and improve quality of life."[60] Beckman's approach to educating chemists and chemical engineers to be system thinkers and entrepreneurs distinguishes his leadership on the pathway to sustainability.

An early collaboration with Michael Buckley, an oral surgeon, steered Beckman's research on polymers toward a practical application.[61] In 2006, Beckman took a three-year sabbatical from his academic work to lead a multidisciplinary team to find a better solution for a medical problem. "When I went on entrepreneurial leave, my research program suffered a lot," said Beckman. "On the other hand, I think I'm a different scientist now. I'm not content to have the end simply be a publication. I want my research to serve a purpose. I want it to impact quality of life in a positive way."[62]

Beckman's research led to a commercial product that solved a significant problem for surgeons. Internal stitching or metal staples often cause complications from irritation and reaction to the surgical materials. The problem is particularly difficult in surgeries of the gastrointestinal system, where the tissue is fragile and subject to fluid accumulation requiring post surgical drains until healing closes the cut surfaces. Beckman's team interviewed both surgeons and patients to have a full scope of the problems and the requirements for an optimal surgical outcome. The material needed to have the right viscosity so it would flow and not clog in the applicator but stay where it was placed. The cure time needed to be slow enough to allow the surgeon to adjust adhering surfaces, but quick enough to allow a stable bond to form within an acceptable surgical procedure. The material needed to be less irritating than polymer filament and cause less inflammatory reaction. And the adhesive should reduce the need for drainage from the surgical site during healing.

The result was an adhesive based on urethane and the amino acid lysine, nontoxic materials that would combine with the moisture from the tissue being adhered. The new patented material, "TissuGlu," forms a polymer network with the tissue to hold the surfaces together long enough for the tissue to heal.[63] In the field trials Cohera conducted as part of the Federal Drug Adminstration approval process, the TissuGlu polymer performed well: the material is biodegradable to small molecules that are readily reabsorbed in the body; it is a nonirritant compared to normal saline solution (used as a standard control for tissue irritation); it is ten times stronger than surgical polyfilament used in internal surgical stitching; and has no foreign body reaction in patients and reduces the need for postsurgical drainage by 90 percent.[64] The FDA has approved this material for use, and it is now produced and marketed by Cohera Medical, Inc. for use in gastrointestinal and plastic surgery and other applications.[65] Beckman remains as the founder and senior scientist, but the company has a full management team and is in commercial operation.

Beckman continues to explore the applications of green polymer chemistry for medical and other applications. He offered this perspective on the role of academics in the development of green chemistry solutions:

> Green chemistry presents a design problem for academics. The civil engineers learn about metrics; the mechanical engineers learn about design; the chemical engineers learn chemistry, etc. They are all in increasingly narrow silos with little cross-communication. Also, engineering students are really goal-oriented. They never see an advertisement looking for "green chemistry" for a job except in nonprofit organizations. Chemists are not taught about several challenges. We need to capture the cost/value of the environmental externalities that means air and water and land pollution. It may be difficult for large companies to do this because they have a vested interest in what is already in place. Small companies can adapt more quickly, or can take on an innovation that is revolutionary and thrive on it.[66]

THAR TECHNOLOGIES: GREEN MANUFACTURING

Lalit Chordia, CEO of Thar Industries, personifies the green technology enterprise in Pittsburgh. When Chordia moved to Pittsburgh from Chennai, India, he had received his BS from the India Institute of Technology in Madras in 1980. He earned a PhD in chemical engineering from Carnegie Mellon University in 1985.[67] As part of his doctoral research, he was working on supercritical fluids experiments with the objective of making biodiesel fuel from vegetable oil. As the work showed promise of success, his colleague Ray Houck started talking about forming a company to commercialize the process. They found a National Science Foundation request for proposals out for which this project would qualify for funding—but the proposal was due the next day! So they stayed up all night and in the morning took the proposal by airplane to hand deliver it to the NSF in time for the deadline. The proposal was funded, and Thar Technologies, Inc. was born.[68] Since 2003, Thar Technologies has diversified into three companies: Thar Process, which works to expand supercritical fluid technology into new sectors; Thar Pharmaceuticals, which works to improve the clinical performance and economic benefits of safe, on-the-market drugs; and Thar Energy, which focuses on green energy solutions.[69]

The innovation driving all of these technologies is the use of supercritical carbon dioxide, recovered as a waste byproduct from making ethanol. A supercritical fluid is any substance at a temperature and pressure above its critical point, where distinct liquid and gas phases do not exist. When carbon dioxide is held at the temperature of 31 degrees Celsius and a pressure of 72.8

atmospheres, it assumes the properties of both a liquid and a gas. It can diffuse through solids like a gas, and dissolve materials like a liquid.

In this supercritical state, carbon dioxide can act as a separation medium, replacing petrochemical-based solvents. There are many advantages of separation by means of supercritical carbon dioxide. The process entirely avoids the use of harsh chemical solvents, such as acids and caustics, which would create toxic or hazardous byproducts or waste products. The carbon dioxide that is used is obtained at a relatively low cost from manufacturers who generate it as a waste product, and because the process is a closed loop, all the carbon dioxide is recovered and reused, not released into the atmosphere. Because the carbon dioxide does not remain in residual products, this process avoids the petrochemical contaminants that may be present in pharmaceuticals and nutrition supplements created by processes that use conventional petrochemical-based solvents. Thus they may be consumed safely by humans.[70]

Unlike most chemical manufacturing processes, Thar uses plant-based feedstocks rather than petroleum, coal, or fossil gas. Chordia offers this example: "We make nutra-ceuticals and food supplements using algae as the feedstock. By growing algae and then extracting the synthesis products, we can have a very pure product free of contaminants, and suitable for inclusion in organic-certified foods." It is basically a biological synthesis process from which specific products are extracted and refined. The Thar process avoids the high temperatures and pressures and harsh chemical conditions used in commercial production, making this a green chemistry production system. The petrochemical solvent traces that are frequent contaminants from commercial pharmaceutical production processes are totally avoided by the Thar synthesis system.

In the energy area, Thar Energy has turned to two different applications for supercritical fluid technology. One is to use carbon dioxide as a heat transfer medium in geothermal earth tube heating. Chordia says, "Because carbon dioxide is a denser heat transfer medium than air, it is more effective in delivering heat or cooling to building space. You can use a one-inch pipe instead of a four-inch pipe, and run it through the floor for radiant heating or cooling."[71] This allows direct-space conditioning without the use of fans and large circulating pumps. The office space of Thar Industries is heated and cooled with this system. Heating and cooling buildings uses about 40 percent of the total energy consumed in the United States. Because geothermal earth tube systems can be retrofitted into many existing buildings, Chordia estimates that about 12 percent of all the greenhouse gas emissions could be reduced by applying carbon-dioxide-based geothermal systems, the largest single application available. Another energy application addresses the enormous use of water in cooling thermal power plants. "By using supercritical carbon dioxide for cooling the steam from power plants, a process called dry cooling, we could reduce the use of both water and energy in power production."[72]

Chordia sees the greatest potential application of supercritical carbon dioxide in the treatment of water through desalination, and plans to focus his efforts on the global water problem. "This is the greatest need for the future. We have tested the process on produced water from hydraulic fracturing, from contaminated water, and seawater." He said, "It could be designed to scale for use in crisis situations such as natural disasters or other crises."[73] Unlike osmosis, which produces one gallon of good water and one gallon of extremely contaminated water for every two gallons treated, the Thar process produces two gallons of good water and sea salt for every two gallons treated. The process oxidizes all the organic contaminants to harmless elements.

WOMEN FOR A HEALTHY ENVIRONMENT: CONSUMER EDUCATION AND ADVOCACY

Consumers are beginning to create market pressure for more environmentally responsible products. When consumers stand together and demand accountability from retailers and manufacturers, results can change the system. More manufacturers are developing products that avoid known endocrine disruptors or toxic materials, but the burden of reading the label and making choices is left to the consumer. Consumers must become vigilant in reading labels, making inquiries to producers and manufacturers, and demanding accountability for safety. There is no scarcity of resources to help people find out about chemicals and components in products used for personal care, cleaning, or food. Some are even available as a phone app, such as the Environmental Working Group's Skin Deep Database that can be used in stores at the point of purchase to scan the bar code for a reading on ingredient safety.[74] But between awareness and action there is a three-step process that begins with becoming aware of the problem of which chemicals are health hazards; taking action to avoid them for personal care products, household cleaning, and gardening; and finally demanding changes in public policy to protect consumers, especially children, and sensitive populations such as pregnant women and the elderly.

Women for a Healthy Environment (WHE) in Pittsburgh has gathered a community to focus on environment and health concerns locally, drawing on the power of larger, more universal organizations for resources. The nonprofit organization educates women about environmental risks and empowers them "so that they can make healthy choices for themselves and their families and advocate for change for a better tomorrow for all."[75] WHE was founded in 2008 as an action from the Women, Environment and Health conferences hosted by Teresa Heinz in Pittsburgh in 2007 and 2008.[76] Each event attracted thousands of women, indicating strong interest among women from a broad cross-section of society across the region. "People want to know how they can manage their lives to protect themselves and their families from exposure to toxic materials in the environment, in homes, and in personal care products," WHE's executive director Michelle Naccarati-Chapkis said.[77]

Moms represent $2.1 trillion in purchasing power and make 85 percent of the purchasing decisions for American households.[78] When this body begins to demand safer and healthier products, manufacturers and product distributors notice. WHE partners with the national organization MomsRising to aggregate consumer power—for example, in the battle to stop the use of long-chain fluorinated compounds for waterproofing, stain resistance, and nonstick applications in household goods. A MomsRising blog post on the compounds reported: "'As a mom, learning about this caused me to ask if I really need products that are stain-resistant, nonstick, or waterproof,' said Joan Blades, cofounder of MomsRising. 'Knowing the potential consequences for my family's health, I will choose to give up some conveniences and product performance. It's just not worth it.'"[79] Naccarati-Chapkis noted: "We do toy testing for lead, especially around the holidays. We do stroller brigades to have moms and babies in strollers to demonstrate at hearings, such as in support of the EPA hearings on carbon emission reduction from power plants." When mothers with children in strollers march as a large group to protest for clean air, it is difficult to ignore.

Consumers have the power to drive change. WHE participates in national coalitions to help people consolidate their voices with other consumers across the country. The Safer Chemicals, Healthy Families coalition represents more than eleven million individuals and includes parents, health professionals, advocates for people with learning and developmental disabilities, reproductive health advocates, environmentalists, and businesses from across the nation. This coalition of diverse groups is united by their common concern about toxic chemicals in homes, places of work, and products used every day.[80] WHE and volunteers have demonstrated in front of stores with the Safer Chemicals Coalition, taken pictures of demonstrators, and put them on social media to create consumer pressure on retailers.

Consumer pressure is effective. In response to consumer demand, large retailers such as WalMart, Target, and several others have required suppliers to remove eight specific chemicals of concern from personal care products.[81] Johnson & Johnson made a commitment to eliminate hazardous ingredients by 2015. The Ashley Furniture Company committed to remove toxic flame-retardant chemicals from its furniture, based on the state of California standard TB 117–2013, as of January 1, 2015. Another significant success involved the removal of BPA from infant baby bottles, initiated in Minnesota and a few other states. Maine required the labeling of toys that contained BPA, leading Hasbro to reformulate its products in spite of opposition to the initiative from the Toy Industry Association and the American Chemical Council.[82] Because manufacturers cannot formulate products for a single state, these kinds of state initiatives result in product safety improvements throughout the nation.

WHE uses the power of a larger coalition to harness resources from the national level, to inform and engage on the local level, and amplify the impact of individual voices. Coordinated action through the national "Mind the Store" campaign has successfully moved retailers to eliminate products containing specific hazardous chemicals. This campaign, a project of the Safer Chemicals, Healthy Families coalition with over 450 organizations participating nationwide, educates consumers and demands action from retailers and government to reduce unsafe chemicals in consumer products.[83] The "Mind the Store" campaign targets big-box retailers, demanding that they look at their sustainability policies and commit to eliminate products that contain the "Hazardous 100" chemicals.[84] An organization like WHE at the local level makes it easy for people to give feedback to manufacturers about eliminating the Hazardous 100. They help people understand the science that identifies specific chemicals of concern because of their effects on child development and endocrine disruption, and provide solutions to empower consumers. Naccarati-Chapkis notes that WHE makes it easy for people to contact their congressmembers by organizing a "day of action" and giving people help with arranging appointments and fact sheets to speak from.

WHE reaches about two thousand people each year through meetings, workshops, and conferences. "People hear the message of taking action to protect ourselves and our families better from a friend, or a trusted person they know,"[85] said Naccarati-Chapkis. In reaching out to women as the chief procurers for the household, she finds that the credibility of the person conveying the information makes a huge difference in the response. "Especially since we are addressing a message that runs counter to the mainstream advertising people may see on TV or in magazines, we need to have the person-to-person connection."[86] Naccarati-Chapkis said that WHE is important in bridging into communities of color: "Our Healthy Homes initiative has expanded through the Train the Trainer program where we identify community leaders in underserved communities of Pittsburgh: Hazelwood, Homewood, Clairton–Mon Valley, Braddock, Lawrenceville. Neighborhood people are trained to give presentations in their own communities to help provide specifically tailored information. This is particularly effective in reaching women of color, especially since many of the products marketed specifically to them contain truly hazardous materials such as formaldehyde."[87]

Many people believe that healthy products are too expensive, or not available to all consumers. While some heavily advertised national brands have created "green" product lines that are more expensive, there are good alternatives that can be much less expensive. WHE holds workshops on handmade, homemade cleaning products and personal care products. In these workshops, people make products from safe ingredients themselves and take them home. WHE tells people that if they change even one product, or take one step, it makes a difference. They look for safer alternatives for cleaning products and comparably priced items to combat the perception that greener is automatically more expensive. As part of the Train the Trainer program, the WHE staff goes with the trainee to shop in the local community to evaluate products available to customers. "People think healthy things are too expensive," noted Naccarati-Chapkis, "but we are talking about our basic health. Everybody knows someone who has cancer, or asthma, or neurological disorders. Scientific studies have proven again and again that the endocrine disruptors are becoming more and more pervasive. We need to stand together and demand changes in the law to protect people, especially infants, pregnant women, and vulnerable populations."[88] Significant changes in the national standards and regulations that could control the production of unsafe chemical products at their source are unlikely in the current political environment; nevertheless, consumers are growing more aware of the issue. It is a significant social justice concern when lax regulatory practices allow corporations to profit at the expense of long-term health effects of the public. The artifice of the current law requiring direct proof of harm from a specific chemical in a specific instance places an unduly high burden of proof on the consumer.

CONCLUSION

Drawing inspiration from nature, green chemists derive solutions based on the experiments of living systems adapted over millions of years. Chemical processes based on green chemistry principles reduce waste and reduce demand on diminishing raw materials, including energy and water. The practice of green chemistry can create alternatives to hazardous substances common in modern materials. By incorporating biological and health parameters to prevent hazards to humans and other living things as a design criterion, the green chemistry processes can displace the endless stream of toxic and hazardous materials that burden the water, air and land. A new regulatory regimen based on preventing contamination, rather than attempting to limit its amounts, would stimulate green chemistry as a path forward to sustainability. The economic realities of health costs and decontamination processes may drive mainstream production in this direction.

Chemicals are part of every aspect of daily life. The long history of modern conveniences has created a society dependent on synthetic materials that often cause immediate harm and frequently cause long-term environmental damage, but the principles of green chemistry applied at the source can significantly reduce the toxic burden on the earth. Adopting the precautionary principle to prevent unsafe chemicals from affecting both humans and other living things in the biosphere can open many more opportunities to manage materials in harmony with nature. Shifting regulatory standards to an approach based on preventing the production of hazardous materials, rather than controlling the amount of unsafe material released, would go a long way toward achieving this goal. Every person can reduce the burden of unsafe chemicals on the Earth by becoming an informed and responsible consumer, and by raising one more voice for precautionary regulations.

The Phipps Conservatory opened to the public in December 1893 and since then has been connecting Pittsburghers to the importance and beauty of our surroundings while advancing the role of the public garden in the fields of education and sustainability.

PHIPPS CONSERVATORY AND BOTANICAL GARDENS

PART III

EMPOWERING CHANGE

We all have a great desire to be able to live in peace and to have environmental sustainability. . . . We can't only blame our governments and corporations for the chemicals that pollute our drinking water, for the violence in our neighborhoods, for the wars that destroy so many lives. It is time for each of us to wake up and take action in our own lives. . . . We have the power to decide the destiny of our planet.

THICH NHAT HANH, "THE BELLS OF MINDFULNESS," IN *SPIRITUAL ECOLOGY: THE CRY OF THE EARTH*, ED. LLEWELLYN VAUGHAN-LEE (POINT REYES, CA: GOLDEN SUFI CENTER, 2013), 27–28

Children learn to plant seeds through the Propel School community garden project.

PROPEL SCHOOLS

7

MOBILIZING AWARENESS TO ACTION

It suddenly became clear. Not only was the livestock industry threatened by a deteriorating environment, but I, my children, my students, my fellow citizens, and my entire country would pay the price. The connection between the symptoms of environmental degradation and their causes—deforestation, devegetation, unsustainable agriculture and soil loss—were self-evident. Something had to be done. We could not just deal with the manifestations of the problems. We had to get to the root causes of those problems.

WANGARI MAATHAI

PRESIDENT JOHN F. KENNEDY'S challenge in 1961 to put a man on the moon opened a new era of aspiration. On July 20, 1969, the Apollo 11 astronauts Neil Armstrong and Buzz Aldrin set foot on the moon only eight years after Kennedy's challenge. This achievement inspired a generation and opened a whole array of new technologies and innovations that permeated the entire American society within a few decades. The success of this venture came from a compelling vision, leadership, and the political will to marshal sufficient resources to accomplish the task. The transition to a sustainable society requires just such a cohesive collaboration.

The pathways toward a sustainable future etch their way through the morass of mainstream practices, in some places more smoothly than in others, and with varying degrees of urgency. All the possibilities struggle against impediments that mire down progress the way thick mud slows the advance of foot soldiers. Applying the ethical principles of taking precaution to protect the living earth and bringing respect and empathy to human interactions opens the way for smoother progress. Concern for the children of the twenty-first century compels cooperation and collaboration across divisions in social, economic, and political factions. A great mobilization is needed to change direction, not merely slow the rate of travel toward disaster.

Climate change and global pollution require a concerted array of actions akin to a mobilization of resources in response to an emergency. On a national level, major transformations have occurred in the past in a very short span of time, often less than a decade. The industrial mobilization that shifted production to make vehicles, machines, and munitions for World War II happened in a span of three years. The rural electrification of America took only five years. The shift from horse and buggy to automobile took only twenty. The shift from regulated communication to unregulated and competitive communications took less than a decade. What is needed in order to mobilize this kind of capability is a clear and urgent motivating force that enables cooperation among competing interests, and a unifying vision to inspire change for the sake of the future. That force can be national security in time of war, market opportunities opened by innovative technology, or collective moral outrage.

Dennis Meadows, who has studied the limits to growth over many decades, writes: "A sustainable world can never be fully realized until it is widely envisioned. The vision must be built up by many people before it is complete and compelling."[1] People today do not have a common vision of what a sustainable world would be like. Apocalyptic views, portrayed in film and story, evoke frightening and unpleasant future conditions with little to inspire optimism and aspiration. However, as communities and businesses begin to take the challenge of climate change more seriously, momentum is building for action. The conditions necessary for sustainability are clear, but the policy solutions to change the current course face daunting challenges, though not from limits of technology or the need for scientific breakthrough innovations. Rather the most serious challenges come from the inertia of historic policies, infrastructure, and customs, and the political power of fossil industry interests.

In the United States, opinion about the entire constellation of issues at the juncture of environment and the economy have become polarized, politicized, and made into abstract caricatures, demonized on both sides.[2] This is the greatest hurdle to addressing issues that require collaboration and broad consensus. Extensive studies of national attitudes toward climate change issues demonstrate that distinct segments of the public hold different opinions about climate change.[3] A Yale Climate Change Communication study found that

> partisan media sources do influence individuals' beliefs about global warming. Specifically, use of conservative media sources such as Fox News and Rush Limbaugh is associated with the belief that global warming is not happening and greater opposition to climate policies, whereas use of non-conservative media such as network TV news, CNN, MSNBC, and NPR is associated with the belief that climate change is happening and greater policy support.[4]

The study described the self-reinforcing effect that results from people making the choice to obtain news and listen to media that align with their own opinions. This self-reinforcing effect is stronger for conservative-leaning opinion holders than for liberal or progressive opinion holders. The study explains this observation by noting, "This may be further exacerbated by conservative media's attempts to insulate their audiences from opposing views—in part, by denouncing the mainstream media (i.e., other news sources) as liberal, biased, and not to be trusted."[5] The authors of the report on the study conclude with a suggestion: "One way forward is the development of communication campaigns focused on specific media outlets and audiences that align climate change and energy insecurity solutions with conservative ideals of limited government."[6] However, in an atmosphere of distrust, there are few effective information campaigns that cross ideological barriers.

The majority of Americans fall in a range of uncertainty about climate change or they question the effectiveness of action. A Gallup poll completed in March 2017 found that "forty-five percent of Americans now say they worry 'a great deal' about global warming, up from 37% a year ago and well above the recent low point of 25% in 2011. The previous high was 41%, recorded in 2007. Another 21% currently say they worry 'a fair amount' about global warming, while 18% worry 'only a little' and 16% worry 'not at all.'"[7] Americans are about equally divided over whether "protection of the environment should be given a priority, even at the risk of limiting the amount of energy supplies—such as oil, gas, and coal—which the U.S. produces" or whether the "development of U.S. energy supplies . . . should be given priority, even if the environment suffers to some extent."[8] A March 2017 Gallup poll finds that this opinion holds: "Americans said the environment is a priority over economic growth by a 50%-to-41% margin. In the thirty years that Gallup has asked this question, Americans have almost always chosen the environment over economic growth as a priority."[9] However, examining the same issue on partisan lines found a wide separation: "Two-thirds of Democrats say the environment should be prioritized higher, while about one-third of Republicans say the same thing."[10]

Clearly, the issue of placing priority on protecting the environment and the associated values associated with clean water, fresh air, fertile ground, and biodiversity of species struggles against immediate concerns over economic security. As the reality of losing basic environmental protections looms as a possibility, people have become more vocal in their concerns, especially about the public health effects of a return to the days of high pollution from unrestricted industry emissions.

ATTITUDES, EXPECTATIONS, AND BEHAVIOR: A PITTSBURGH CASE STUDY

> *Our future depends on the ability of our government to work together for the common good of everyone, rather than those wealthy enough to fund their campaigns and help them get reelected. Our children, grandchildren, and future generations deserve the same right we had to inherit a habitable planet, and I hope we realize this in time to really make a difference.*
>
> **ANDREW SLAUGHTER, UNIVERSITY OF PITTSBURGH STUDENT**

In an effort to address the limited question of how people in Pittsburgh view their role in climate change, I engaged Campos Research Strategy to conduct a focus group study.[11] The study screened participants to select people who were not clearly aligned with a specific position, but were undecided or open to ideas on climate change. The participants were men and women of ages ranging from thirty to sixty, of a variety of races, and with at least a high school education. The major findings indicate a starting point for public perceptions about action on climate change.

The respondents did not make a connection between burning fossil fuels and climate change. They described climate change as "melting glaciers, extreme weather/catastrophic events; oceans heating up, changing weather patterns."[12] For the most part, respondents saw the changes in climate as a global issue, but did not connect their own behavior, as individuals or as members of a region, or country, as having a direct impact. Respondents had little awareness or knowledge about fossil fuels—what they are, how they are being used, and how they are developed. Using a frame of reference from the days of steel and heavy manufacturing, for the most part, respondents felt that the Pittsburgh region has made great strides in improving the environment. They see blue sky and do not observe heavy smoke in the air, so the dominant opinions were "we are doing pretty well"; "we are doing our part"; "things are getting better." Thus, the level of concern about climate change and its impact does not present a priority for them to change their behavior.

The participants had very little awareness or knowledge about "clean power." They were unsure of what renewable energy sources are or how renewable energy can be delivered through the current system and infrastructure. However, there was a lot of concern about utility companies offering clean power due to an environment of distrust based on experiences with deregulated telecommunications marketing and other "bait-and-switch" offerings in the marketplace. People were not sure how green energy could be delivered in the same wires that deliver nonrenewable energy. Although respondents did believe that organic produce contains fewer contaminants such as pesticides and other chemicals, they distrusted "green marketing" in general and showed little confidence that "green products" indicated any real difference from conventional products, except for green products having a higher price.

The respondents believed, at the big-picture level, that society as a whole needs to address climate change. They agreed that "what we do today can impact future generations" is important in the same way that everyone agrees "we all want world peace." This is consistent with broader statewide surveys showing that nearly three in four (72 percent) Pennsylvania voters favor the EPA's proposed regulations to limit the amount of carbon pollution that power plants can release into the air, while just 23 percent oppose the EPA's regulation of carbon pollution.[13] However, as a motivation for behavior change, this is too grand, not specific enough, and does not communicate a specific benefit to the current generation at the individual level.

Respondents were highly concerned about the impact of climate regulation on workers and job loss. When given examples about energy conservation or efficiency behavior changes that could displace the need for coal power plants, respondents' immediate reaction was concern for job losses and the negative impact on individuals and families of the power plant workers and coal miners. This may be the result of the history of this region with the tremendous loss of jobs due to the steel industry's decline in the 1980s. Participants questioned the benefit to the environment at the expense of people's lives. It was difficult for the group to prioritize local or individual actions within the bigger picture. Respondents did not make a connection between burning fossil fuels and their own health.

Regarding motivation for individuals to take action, *respondents would be motivated by energy changes that produce monetary savings.* However, they would need evidence of claims of savings in the form of facts from credible sources. Credible sources included most importantly a trusted person they know—a friend, a pastor, or close business associate. They suggested having visual presentations to communicate quickly and easily, like social media updates or infomercials, to demonstrate the impact of actions and show the results. Such information should appear on mainstream programs such as commercials during football games.

The participants believed that energy conservation and efficiency programs are important. Nevertheless, "you need to make energy savings fun"—communitywide programs of competition or school programs would be helpful for motivating participation. They thought it was important to get kids involved. One participant said, "It's the younger generation that will change their behavior if we teach them now."

The respondents noted a number of impediments to making changes in behavior that would address climate change. They noted that people are stubborn and set in the ways of old habits. They are unlikely to change behavior even in small ways without outside encouragement. People do not like to feel they are the only one making changes, especially if they cannot see a direct impact of their actions. Seeing concrete results and receiving recognition for making changes seemed important to people.

There was a prevailing sentiment that if the change was important for society, then society should pay for them to make the change. As one man stated, "If it is going to cost me, who is going to pay for it?" They did not feel compelled to make changes to address climate until they could see a direct negative effect themselves. Participants believed that the Pittsburgh region seems to be an example of an area that is improving its quality of life and environment compared to its history of steel mills and heavy industry. Thus, respondents felt, "we must be doing our part in improving the environment and climate." The historical comparison allows people to feel as if they are "off the hook" in feeling some sense of responsibility for climate change and the quality of the environment.

SHIFTING CULTURE AND VALUES

> *I really hope that my children will not have to live on a planet where seeing nature is a thing of the past and only get to see a concrete, or trash, jungle. North America, and the rest of the world, has so much natural beauty to offer and I do not want to see any more of it taken away or destroyed.*
>
> **HOPE COMDEN, UNIVERSITY OF PITTSBURGH STUDENT**

One major area of uncertainty for Americans has to do with the lack of knowledge or understanding of how alternatives would affect either the climate situation or their own lives. In 2014 I convened a roundtable workshop for the Institute for Green Science at Carnegie Mellon University, drawing together twenty people from gas and electric utility companies, solar power and windpower manufacturers and distributors, conservation and efficiency service providers, and academic researchers. In this roundtable, we discussed the actions that would be necessary to accelerate the use of renewable energy systems as much as possible. After covering the infrastructure issues and economic considerations involved in such a shift (discussed in chapter 6), we turned to cultural or values-based issues that pose a challenge and also suggest opportunities for shifting social values toward an ethic of conservation and resource preservation.

Most of the roundtable participants believed that it will take a generation to shift cultural priorities to a different set of values. Shaping that shift will require education and building awareness among children. There is already a tipping point among people in the millennial generation who have expectations that the world will be different. Students are very enthusiastic about renewable energy, but two centuries of an economy based on fossil fuels will not change overnight. The political system is reactive, not proactive. Even when surveys show public support for renewable energy to be at 70 percent and up, the political system still stays in the fossil arena.[14] There was a sense among the roundtable participants that cultural change needs to happen sooner than 2021.

Roundtable participants believed that while a global perspective is central for understanding climate change, resource conservation is not currently a priority as a value. Participants noted that people have difficulty in feeling empathy or a sense of personal urgency for people in other places. Cultural conditions make change difficult; however, instilling a culture of care for the environment, for community, and for the needs of future generations is clearly necessary. Climate change presents a complex global issue—developing countries have expectations; sustainable energy awareness has to penetrate to policy change. Even when a new ethic of environmental custody is defined and implemented, it will take thirty to forty years to take effect.

Extreme climate events of the last few years have sensitized the public to climate change effects, but as yet climate events are difficult to connect to current actions in an immediate cause/effect relationship. As they become more frequent, the cost of this situation will compel action. Disaster presents a catalyst for change, but having a system where multiple sectors work together will take time. The Pittsburgh region, for example, has 136 towns that are very fragmented in jurisdiction with engrained habits of maintaining control. Individual cities and towns are making efforts, and there the boundaries will be tested and new solutions will form.

The roundtable ended with a consensus on education. Participants agreed that the general public is not well informed about the energy system or how it relates to large issues like air pollution, climate change, or the economy. Wider public understanding is one part of the answer, but it needs a well-thought-out strategy to be effective. Creating a new culture of conservation requires leadership and broad messaging. It was suggested, for example, that outreach and education in the mainstream can include "This Is Your Future" PSA spots creating positive images of keeping the earth alive through renewable energy.

There was a high sense of frustration with the limited resources devoted to an action-oriented education effort. One participant noted, "You can give people energy kits and show them how to use them, and come back and still find they have not done the actions. We need a really different approach. Community-based and really simple works."[15]

Visualizing a sustainable future based on renewable and sustainable resource use is difficult when such images are not common in the mainstream media. Social marketing and behavioral changes need to be based on social norms. Making energy efficiency and renewable resources the new normal needs positive feedback to make such practices regular behavior. Renewable systems need to be in place among people who can see them. As one roundtable participant commented, "We need to work with the culture we have and create multiple places where people encounter change—workplace, worship spaces, recreational areas—engage where life is important and where people gather in schools, town halls, libraries, offices."[16]

MOVING FROM AWARENESS TO ACTION: PITTSBURGH PATHWAYS

In the face of a resistant attitude toward change, there are several initiatives making progress in communities around Pittsburgh. Their experiences in motivating participation reveal the elements for success. Building awareness is a necessary first step, and critical to the success of any broad societal change. However, even with wide awareness of a problem, people and businesses do not move or take action without specific motivation, recognition, or rewards. The process for combining all of these elements into a functioning program can be done by prescriptive regulations—such as the Renewable Energy Portfolio standard—or in more creative ways. The challenge is to make such initiatives more broadly applied and adopted throughout the community and throughout the country. Among the many initiatives underway in Pittsburgh, several examples offer insight into ways individuals, businesses, institutions, and communities empower others to move from awareness to action. An ethic of concern for the well-being of future generations and preserving the environment runs as a quiet thread through these efforts to overcome the barriers to change.

THE BREATHE PROJECT

Taking action to address climate change comes from different directions and different perspectives. Although the US population is deeply divided on climate change, especially in considering the role of humans in causing it or having an effect on improving it, there is less division or disagreement about health. The Allegheny Conference on Community Development, a coalition of Pittsburgh business, civic, and community leaders, dates back to the post–World War II era, when it was formed in part to improve the region's poor air quality, which was seen as an impediment to the business climate and quality of life. Although the details of how to clean up the air still runs counter to the Allegheny Conference's "full portfolio" energy strategy of the region today,[17] placing the focus on clean air by reducing emissions that cause asthma and respiratory distress creates a broader base for agreement.

Environmental contaminants rarely precipitate a health problem in isolation, but often act as complicating risk factors for a large number of adverse health outcomes. For example, ambient air quality has an effect on people's willingness to exercise outdoors, especially if contaminants prevalent in the air have noxious odors. All people are not affected in the same way by exposure to pollutants, and the ambient conditions expose people to complex mixtures of contaminants that are difficult to segregate.

Some statistical correlations between pollution and diseases are well documented. The Centers for Disease Control and Prevention report that 7 percent of adults and 8.3 percent of children suffer from asthma nationwide; in areas with higher air pollution, the levels rise to 15 to 25 percent of the adult population.[18] Children, the elderly, or persons with underlying chronic diseases are at higher risk from pollution-induced asthma, especially the small particulates associated with power plant emissions.[19] Mercury, also emitted from coal- and oil-fired power plants, is a known neurotoxin, especially for unborn children. Controlling mercury emissions, as required by the Mercury and Air Toxics Standards of the Environmental Protection Agency, prevents up to 11,000 premature deaths, 4,700 heart attacks, and 130,000 asthma attacks every year.[20]

People who live in areas with cleaner air and water tend to be healthier, and everyone can play a part in making a healthier environment—energy companies, large industries, and individuals as well as environmental organizations. It is a space for common ground. The Heinz Endowments initiated a project in 2014 to gather a broad coalition of companies, organizations, and government entities to work together on Pittsburgh's air problem. Their call to action states: "Despite improvements over the past few decades, our region's air still ranks among the worst in the nation, exacerbating asthma and causing other serious health problems in our communities such as higher rates of heart disease and lung cancer. But there are solutions. If you care about the air you breathe, we invite you to join us. Our lives—and way of life—depend on clean air."[21]

The Breathe Project is a coalition of residents, businesses, government, and many other groups in southwestern Pennsylvania working together to clean up the air for the health of families and the economy. The project uses emissions data as the common metric for evaluating program success among efforts of all participating organizations. The emissions of greatest concern are fine particulate matter, solid particles of 2.5 microns in dimension (PM-2.5),[22] benzene and ozone, measurements of which in the Allegheny region frequently exceed national standards for air quality. The program has achieved 16.5 tons per year in reductions of particulate matter in the air from all partners in the project.[23]

Air monitoring and modeling data allow people to track air quality in real time from the Allegheny County Health Department website.[24] The project also has input on community planning initiatives in the city of Pittsburgh to focus public attention on creating walkable centers and pedestrian amenities to encourage nonmotorized travel options that have no air impact. It is important for pedestrians and cyclists to have travel spaces safe from exhaust and emissions from other vehicles. The Breathe Project includes mobile monitoring on bicycles by citizen participants in gathering data, and the effort to make the streets more bicycle-friendly and safer for pedestrians is tied to this process.

CONSERVATION CONSULTANTS, INC.

Individuals are motivated to make changes to their energy consumption when they see cost savings, increased value in their property, and increased comfort and performance of their home energy system. Conserving resources becomes a quality of life issue, which has broader acceptance and a less-complicated justification. Conservation Consultants, Inc. (CCI), founded in 1978 as a private nonprofit organization, promotes responsible energy and resource use in homes and buildings. CCI helps people improve their personal comfort, health, and safety, and increase awareness of the positive benefits to the economy and national security afforded by energy efficiency and renewable energy.

The organization achieves its mission through education and identifying efficiency improvements for homeowners, renters, and communities. Funded by electric and gas utilities through the Pennsylvania Act 129, CCI helps customers understand how they can save money by reducing utility bills from making energy efficiency and conservation improvements.[25] They train professional energy auditors and educate consumers about how to protect and preserve the environment by reducing the use of natural resources.

Jeaneen Zappa, executive director of CCI, noted that the process of convincing people to take action on their own as an outcome from education programs is a complex dynamic. In most cases, education and awareness alone do not drive a change. Zappa said, "It is really a challenge to move people from awareness to action. Even when they see that changes will save money, people are reluctant to make permanent adjustments in behavior such as turning off lights, or using natural light for indoor tasks."[26] Even with personal energy audits, people need prompting, and personal hands-on follow up to guide them through the transition from knowing about options to actually taking action. "You have to walk them through the implementation and provide hands-on assistance for things like simple weatherization, or caulking leaks," Zappa said.[27] She explained, "Getting the envelope of the house as tight as possible is just so unsexy! The message is all about comfort, home health and indoor air pollution, and safety. We talk about helping people get the most for their investment in their home with respect to efficiency, health and safety. We are successful because our staff builds a connection with our customers. Our long history in the community gives people a sense of trust that we can build on to help motivate them to take action."[28]

CCI's mission depends upon strong partnerships with public utilities, government agencies, foundations, other nonprofit organizations, and for-profit businesses. Each year CCI's partnerships help the organization to improve the lives of citizens by providing over 5,000 energy audits of homes and buildings, education services to over 1,000 students and young adults, and 1,800 water savings kits across the Commonwealth.

Zappa noted, "There is a huge difference between how things are handled between the gas utilities and the electric utilities. We need to focus only on the electric uses for audits done for electric utilities, because the ratepayers are paying for the audits. There needs to be much better integration, and there should be a way to share information across the utilities."[29] The information about consumers' use of gas and electricity is protected for customer privacy and also for utility providers' confidence that their information is not being given to competitors. In some cities in other states, the approach is different—for example, in Seattle, there is a "Green Button" that allows a customer to consolidate all of their gas, electric, and water utility data in one place.

In terms of changes that would alleviate impediments to efficiency, a requirement to bring houses up to code for energy efficiency at the time of sale would make a huge difference in the perceived value of energy efficiency. "We can increase both awareness and action if we routinely disclose the energy efficiency and performance of buildings at the point of sale or rental," said Zappa. "This would be a huge help in moving the market value to include energy efficiency into the realm of consumer awareness and preference. It is not allowed by state law for municipalities or counties to require additions to the real estate transaction requirements, but a voluntary disclosure would be a great start."[30]

To make this a positive action rather than a disincentive, especially for owners of older homes, people would need to have the energy efficiency tax credit as an incentive. According to Zappa, "This kind of work is local, ongoing, and essential for incrementally improving the overall performance of the buildings in the community. If this kind of requirement were in place, it would generate demand for semiskilled labor to meet efficiency codes."[31]

SUSTAINABLE PITTSBURGH'S GREEN WORKPLACE CHALLENGE

The Green Workplace Challenge (GWC) initiative, seeded with a grant from the Heinz Endowments, began in 2011 in response to the Pittsburgh Climate Action Plan goals to reduce climate emissions by 20 percent, to below 2003 levels by the year 2023.[32] The GWC gives businesses, nonprofits, municipalities, colleges/universities, and K-12 schools the opportunity to include sustainability-based activities in an approachable competition that provides tools, information, and guidance necessary to reduce costs, improve performance, and increase long-term environmental sustainability. By constructing a competition with specific metrics for measuring the effect of actions taken and a broadly publicized recognition program, the GWC—now in its fourth thirteen-month cycle—has made a significant impact. The first three cycles have engaged more than two hundred participants, completed over 5,800 actions that saved 111 million kilowatt-hours of electricity, saved 124.5 million gallons of water, and reduced the carbon dioxide equivalent of 25,485 metric tons.[33]

The GWC steps beyond national trendsetters in enabling participants to actively track and receive credit for their verified green actions in a secure environment. "It was important to find a way to verify the effect of actions taken without compromising the confidentiality of business practices among competitors," said Matt Mehalik, program manager for Sustainable Pittsburgh, who managed the GWC until 2016.[34] Because the program involves the verification of potentially sensitive information, all data used in the competition remains confidential and secure. Only broad characterizations of competition performance (in the form of competition "points") and aggregated savings across all participants are public. Participants in the competition can track their current point totals using the GWC leader board, which is updated continuously for most actions and at least monthly for energy, water, waste, and transportation actions and displayed on the competition's website.[35]

Mehalik noted, "It is important to recognize that every little bit counts. Every action can make a difference. It is the effort and the engagement in a process that matters." He described the experience of a woman in the accounting department of one of the competing companies: "She told me that by participating in the Green Workplace Challenge her job became fresh and exciting because she was empowered to stamp out waste. It gave her a sense of purpose and meaningful contribution."[36] This process of empowering people to take action in their own sphere of influence, knowing that they are contributing to a larger purpose by working together, will change the system. Having broad public recognition as a significant part of the competition creates peer pressure for sustainability actions, and increases the culture of continuous improvement and positive change. With broader recognition and motivation through peer pressure, sustainability becomes the new normal.

GTECH STRATEGIES: GROWTH THROUGH ENERGY AND COMMUNITY HEALTH

The city of Pittsburgh has over fourteen thousand lots devoid of any kind of structure and varying in size from twenty-six square feet to 634 acres. Abandoned properties and ghost landmarks like former elementary schools are ubiquitous. The mayor's office, nonprofit organizations, and neighborhood associations have attempted to tackle issue of vacant property with limited success. However, in 2007, a graduate student project on vacant land in Pittsburgh shone an outside perspective on this problem.[37] One of those students was Colorado native Andrew Butcher, who cofounded GTECH Strategies (Growth Through Energy and Community Health), a nonprofit organization that cultivates the unrealized potential of people and places to improve the economic, social, and environmental health of communities.[38]

GTECH found opportunities in waste by reclaiming Pittsburgh brownfields, growing sunflowers, and making biofuel from waste restaurant oil.[39] Sunflowers are particularly effective in reclaiming brownfields because the plants take up heavy metals and other contaminants from the former industrial sites to improve the ground, and their seeds can be processed easily to make biofuel. By farming one of Pittsburgh's largest brownfield sites—by hand—GTECH piloted the concept of growing biofuel crops to remediate land and produce biofuel feedstock. The approach was not just successful in catalyzing change but was effective in bringing together diverse and complementary partnerships.

The program, called ReFuel Pgh, expanded the initial sunflower field to several other communities in the following years. In 2013 GTECH partnered with Greenlight Biofuels to add sunflower oil to reclaimed cooking oil to make biodiesel. In a company blog post about the launch of the program, Butcher explains, "Not only does this new partnership expand the reach and capacity of ReFuel Pgh, it allows restaurants of all sizes, institutions, stadiums, and national chains the ability to achieve a community benefit and an environmental win with their waste stream. How could you not be excited about a win-win-win scenario?"[40] In October 2014, GTECH stopped its direct involvement in this program, leaving Greenlight Biofuels to operate its successful enterprise.

GTECH has adapted its approach to a community scale. The company's methodology is the core component of their programming: "We believe that investigation is an ideal starting point to address systematic problems that communities face."[41] The process begins by engaging a neighborhood or community in an investigation of what problems and opportunities are present around vacant land, energy efficiency, and wasted resources. GTECH provides resources and creates processes to allow all stakeholders to participate and to see best practices from other places for ideas. An action plan emerges to turn environmental liabilities into assets. The company views itself as a catalyst to connect important partners and to help bring necessary resources to the project. GTECH considers a project successful when the community takes ownership for ongoing implementation of their vision.

GTECH has cleaned up fifty-six acres of vacant land and installed twenty-three permanent projects, from community gardens to chess tables, through its ReClaim program.[42] After almost ten years of operating the company, Butcher is still "compelled by the prominence of vacant land in Pittsburgh, which perpetuates a downward cycle of poverty."[43] He believes that by strategically using vacant land, such properties can turn into an opportunity to

Members of Operation Better Block (Von, Lydia, Ryan, and Elijah) partner with GTECH to implement four new greening projects in Homewood.
GTECH STRATEGIES

benefit more people.[44] Having grown up in a household full of passionate women, including the Iditarod Race champion Susan Butcher, Butcher understands the importance of listening to others and of considering other people.[45] He also believes it's critical to "invest in people" through programs like ReClaim, which provides GTECH community ambassadors with professional development and training and connects them with designers, policy specialists, and landscape architects.

Butcher noted, "Through educating and empowering individuals in the community, GTECH has had a much greater impact than if the GTECH team had gone into neighborhoods and tried to reach people themselves."[46] This community-centered approach allows the people who live in the community to take full ownership of projects. It has created a cohort of over one hundred ambassadors, who have the proven ability to manage community projects and the skills and experience to help others create positive community change.

Rhonda Sears is one of the GTECH ambassadors. A master gardener, community advocate, and voracious learner, Sears believes that her work with GTECH "helps people better their lives and to improve her neighborhood."[47] As a GTECH ReEnergize Ambassador, Sears helps people in the community to reduce home energy consumption and to increase the demand for energy efficiency improvements. "In this role, trust is critical," she noted. "People are worried about being scammed."[48] Sears's credibility in her community and her ability to draw out community needs helps her to introduce energy saving information in a nonthreatening and relevant manner. She adds that her connection to the community "makes me work a little harder to share my knowledge and resources."[49] During her first year as an ambassador, for instance, Sears spoke with 340 of her neighbors about energy efficiency, and her work "kept more than 250,000 pounds of carbon dioxide out of our air."[50] Sears believes that the opportunity to share and vet ideas with other ambassadors was critical to making her ReEnergize efforts more effective.[51]

GTECH has a dynamic feel, a sense of picking up the best of what a community has to offer, and connecting the pieces in new ways. Butcher glows with enthusiasm when he talks about his company. "When we start to engage with a community, we can find ways to turn problems into the foundation of solutions. It is an interactive process, and no one person ever has all the answers. All of us together are smarter than any one of us alone."[52] He hopes that programs like ReEnergize help harness the passion of people, create jobs in energy efficiency, and inject energy and resources into forgotten neighborhoods, so that eventually vacant land becomes a resource rather than a liability.[53]

CONCLUSION

> *Here and there awareness is growing that man, far from being overlord of creation, is part of nature, subject to the same cosmic forces that control all other life. Man's future welfare and probably even his survival depend upon his learning to live in harmony, rather than in combat, with those forces.*
>
> **RACHEL CARSON**

Half a century ago, seeing the blue, green, and white swirled marble of Earth from the perspective of space marked a paradigm shift in human expectations. For some, this represented the first step on the conquest of yet another frontier. Others saw the earth from the distance of space and recognized its fragility. Astronaut Frank Borman of Apollo 8 said, "I think the one overwhelming emotion that we had was when we saw the earth rising in the distance over the lunar landscape. . . . It makes us realize that we all do exist on one small globe. For from 230,000 miles away it really is a small planet."[54]

Societal transitions rarely happen quickly. They emerge from an accumulation of smaller movements over twenty or thirty years until, at last, sufficient pressure for change has accumulated to overcome the inertia of common practice. Unlike past transitions driven primarily by innovations in technology, this twenty-first-century transition must be compelled by changes in the global environment that potentially threaten life on Earth. The transition to a sustainable society requires a new paradigm for human enterprise that values environmental preservation and social justice above short-term economic gain. It will require a change in consciousness and values where people recognize both the fragility of our planet's complex living systems, as Borman noted, and the impact human lifestyle and demands have on its viability, as Maathai observed. In this transition, a change in the ways human technology and inventiveness are applied must be driven by a passion for preserving the living earth and reducing human impact on its living systems. Carson's ethic of living in harmony with nature's laws provides a prescription for a change in direction. The transition depends critically on people's understanding that we are part of a living planet, and that we will thrive only by preserving the health of all living things. When communities come together to address common problems, the solutions rise from within the society and take lasting effect. A culture of caring for the earth and for each other gradually emerges into the mainstream of society.

Celeta Hickman proudly shows off the Ujamaa Collective's fifteen-acre farm in the Hill District.

UJAMAA COLLECTIVE

8

THE NEW ECONOMY

VALUE BEYOND PRICE

We are living in a time of crisis; we see it in the environment, but above all we see it in men and women. This much is certain—the human person is in danger today, hence the urgent need for human ecology! And the peril is grave, because the cause of the problem is not superficial but deeply rooted. It is not merely a question of economics but of ethics and anthropology.

POPE FRANCIS

A PARKING LOT outside a stadium is littered with debris after a big event. Hundreds of people have discarded plastic beverage bottles, wrappers, food containers, plastic bags, and cigarette butts. A sudden summer thunderstorm washes much of this material into the storm drains and overflows into the nearby Ohio River, from which it then floats down to the Mississippi River and into the Gulf of Mexico to drift into the ocean. Plastic debris enters the oceans from coastal areas and river deposits at the rate of 4.8 to 12.7 million metric tons per year.[1] Fragmented by the mechanical action of the ocean, the debris injures ocean creatures from whales to plankton. A gyre of plastic has accumulated in the Pacific Ocean measuring miles wide and a half a mile thick. Ocean birds have been found full of plastic pieces, and millions of sea animals are tangled and killed in this waste of modern civilization.

The systems we use for producing energy, food, and materials share a common theme of waste and inefficiency. Public awareness of planet earth's resource limitations is and has been growing steadily; however, critical resource limitations come not from scarcity of raw materials, such as energy supplies, but rather from the failure to manage the balance of resources, especially the wastes. Using the oceans, atmosphere, and land as waste repositories now takes a terrible toll on the natural functions these would serve. Controlling global pollution and climate change requires a new paradigm for business to reflect both the irreplaceable value of ecosystem services and the cost of using resources without regenerating or replacing them. For example, wood in the form of lumber costs four dollars a linear foot, but a forest is priceless in value. Movement toward this new paradigm can take place if it is driven by three major motivations: moving from a linear to a circular flow of materials in the economy; facilitating the convergence of technologies for global communications and distributed energy systems; and equity and justice in meeting rising global expectations.

SHIFTING FROM RESOURCE EXPLOITATION TO SUSTAINABLE USE

> *Society is faced with a profound dilemma. To resist growth is to risk economic and social collapse. To pursue it relentlessly is to endanger the ecosystems on which we depend for long-term survival.*
>
> **TIM JACKSON**

For the sake of preserving a stable living system on earth for current and future generations, limits to human actions causing climate change and global pollution must be put in place. However, the assumption of unlimited freedom to use the resources of the earth for economic gain lies at the heart of modern global economic enterprise. Earth's resources and creatures are valued as they are perceived to have use for humans, and each country, institution, and individual tends to act in its own immediate best interest, even when all agree that collective action must be taken. Significantly, the modern industrial economy is based on extracting resources from the earth, manufacturing goods with the investment of infrastructure and labor, and selling materials to consumers, who use and discard most of the products. The economy moves material from raw resources to trash, placing value only on the component that sells for a price in the market. The costs to the resource base, the natural capital from living ecosystems, do not count explicitly in the calculation of gross domestic product, for example.

Examining the corporate mission and principles for one of the largest multinational energy companies, ExxonMobil, reveals the corporate focus on expanding the use of petroleum while assuming energy use advances development and human society: "ExxonMobil is committed to being the world's premier petroleum and petrochemical company. To that end, we must continuously achieve superior financial and operating results while simultaneously adhering to high ethical standards."[2] The principles that they articulate as driving this mission explain what these corporate ethical standards are: earning superior returns for shareholders, providing consumer products at competitive prices, and adhering to applicable safety and environmental laws.[3] In the context of the corporate mission, then, adhering to applicable laws leads to a strong corporate interest in monitoring, shaping, and influencing laws that will affect their corporate objectives.[4] Companies in the petroleum, natural gas, and coal industries have resisted environmental regulations consistently since the 1970s, as such regulations are assumed to require greater expense and less efficient operations, cutting into their profits.

This profit priority, which has stood as a standard for business since Victorian times, diminishes emphasis on wider societal consequences resulting from the combustion of fossil fuels. As global environmental conditions have deteriorated, largely from human actions, the need for a change has received broader recognition. And while international climate change responses focus on how the shift away from burning fossil fuels will occur, the social and political adjustments necessary to shift the economy from a fossil-fueled basis to a renewable and sustainable one require a broader debate on the topic of how people use and share the resources of the earth. Climate change thus presents a growing series of challenges for technology and innovation but also creates the need for a new value system based on the vital necessity of preserving the living earth. Nonhuman living plants and creatures have a right to exist, and human existence itself depends on their thriving.

The concept of sustainability includes the balance of resource use with equity and fairness across all populations and future generations. In 1983 the United Nations established a World Commission on Environment and Development with the charge to define a global agenda for change. Gro Harlem Brundtland, chair of the commission, noted the gravity of this undertaking: "Many critical survival issues are related to uneven development, poverty, and population growth. They all place unprecedented pressures on the planet's lands, waters, forests, and other natural resources, not least in the developing countries. The downward spiral of poverty and environmental degradation is a waste of opportunities and of resources. In particular, it is a waste of human resources."[5]

The Brundtland Commission Report of 1987 defined the concept of sustainability in the context of this weighty undertaking, and although many variations of a definition have emerged since then, the Brundtland definition remains a well-accepted basis for defining sustainability as "meeting the needs of the current generation without compromising the capability of future generations to meet their own needs."[6] The concept incorporates the idea of sufficiency without excess; preservation, replenishment, and conservation of earth's resources; and fairness in resource allocation. As a foundational principle, sustainability requires that natural resources be regenerated and restored rather than depleted, an essential element for sustaining nonhuman life on earth and its biodiversity. A sustainable balance equally supports the cultural and social stability of society—to have a sustainable civilization, the demands of the population cannot exceed the capacity of the earth's resources.

Current international economic activity has been described as an intrinsically unsustainable system. Paul Hawken, Amory Lovins, and L. Hunter Lovins have been working for decades to establish a new way of doing business in the twenty-first century. They argue that the basic premise is flawed: "Capitalism, as practiced, is a financially profitable, non-sustainable aberration in human development. What might be called 'industrial capitalism' does not fully conform to its own accounting principles. It liquidates its capital and calls it income. It neglects to assign any value to the largest stocks of capital it employs—the natural resources and living systems, as well as the social and cultural systems that are the basis of human capital."[7] Their goal for a sustainable business model in the twenty-first century sets out a paradigm in which nature and people are fully valued, but not monetized.[8]

As climate change and global pollution receive greater recognition, the concept of corporate social responsibility, including sustainability, has increased in influence. Many corporations and institutions now include sustainability among their objectives.[9] The primary focus on profit has expanded to include other societal values within corporate social responsibility principles, as in "the triple bottom line—people, planet, profit."[10] The Pittsburgh P4 plan for sustainability and resilience also includes place, reflecting local conditions and values.[11]

Proposals for realigning the economic framework within which corporations operate, to include value for human capital (labor, intelligence, culture, and organization) as well as natural capital (natural resources, living systems, and ecosystem services), have been articulated and have begun to penetrate mainstream business practices;[12] however, the current economic system based on capital investment and reward from the cycle of resource extraction, manufacture, consumption, and disposal of goods poses strong resistance to change.[13] Conflicting objectives surround the many issues related to the distribution of costs and effects of making such a major change in the economic foundation of industrialized society. The necessary transitions from despoiling to preserving the resources of the earth will not happen under the market-driven systems that created most of the problems. But approaching the issue from the perspective of moral responsibility offers a broader prospect for success in achieving change.

Moving to an economy based on renewable energy, restorative agriculture, and green chemistry addresses several ethical concerns simultaneously. First, moving to systems that inherently conserve and regenerate resources helps control—and may ultimately reverse—environmental degradation. Using practices that conserve and regenerate resources helps to restore the natural systems of the living earth and brings human consumption patterns within range of the carrying capacity of the earth. This shift can come about through recognizing the obligation of each generation to preserve the natural resources of the earth for the next generation.

Second, an economy based on circular resource use can provide meaningful work for people within the communities where they live. The processes for recapturing, reusing, and restoring resources creates a tier of employment opportunities available as soon as the value of recovered material can be reflected in the economy. Making such jobs valued as part of the mainstream economy can displace unhealthy, dangerous, and abusive practices such as having children in third-world countries labor over Bunsen burners in electronic waste dumps to recapture precious metals.[14]

Third, a sustainable economy extends the capacity for more people around the world to enjoy improved living conditions, and responds to development expectations with less impact on the natural world. Resilience and security can result from investing in the infrastructure necessary for renewable and sustainable practices. The waste of human resources, abandoned land, and neglected infrastructure can be reversed by community commitments to equity and justice at a local level. Resources brought to rebuild and reinvest can be drawn through community initiatives to generate more sustainable solutions. This is an especially fruitful pursuit in reshaping the food system away from commercial monoculture, much devoted to feeding livestock, in favor of a more disaggregated system of regenerative farming that includes urban farms.

A NEW ECONOMIC MODEL

The sun and wind are free fuels and universally available. What is costly is the technology and mechanisms for harnessing the energy value of these resources. In this regard, fossil fuels are similar. The resource as it lies in the ground is free, part of the earth's crust. But a system for ownership and control of fossil resources has been created as property rights. The extraction and exchange of these materials has allowed businesses to accrue value based on these transactions. For example, the amount of natural gas and oil determined as "technologically recoverable" varies as a function of the price. When the price is higher, there is more reserve determined to be recoverable.[15] Over the last fifty years, wealth has consolidated in the hands of those who have developed the fossil resources of the earth as fuels, and they hold the power to control the economic fortunes of people all over the world who consider fossil fuel requirements to be absolute necessities for civilization.

As frustration with the wealth concentration in the top 1 percent of the economy increases, a wave of community wealth-building institutions has begun to swell across the country. People are joining together through a variety of forums such as public-, community-, or employee-owned businesses to meet local needs and thus regain a sense of democratic control. Community development corporations, community banks, social enterprises, community land trusts, and employee-owned business and cooperatives emerge as the instruments for building community wealth.[16] Worker-owned businesses include manufacturers, retailers, and a number of nonprofit organizations. Community Development Corporations (CDCs) can now be found in nearly every major city across the United States. Once limited to redeveloping blighted areas following urban riots or rural neglect, these CDCs produced over 1.6 million units of affordable housing nationwide over the last two decades.[17] All of these institutions pool capital in ways that build wealth, create living-wage jobs, and anchor those jobs in communities.[18]

A major initiative that can shape a more sustainable way forward involves redefining the structure of the economy away from a model driven by corporate profit motive to one based on corporations operating for social benefit. One such model, the New-Economy Movement, has several manifestations, all gaining momentum in different ways. The movement seeks an economy that is increasingly green and socially responsible, based on rethinking the nature of ownership and the growth paradigm that guides conventional policies.[19] Gar Alperovitz, one of the leaders of the New-Economy Movement, says, "Over the past few decades, a deepening sense of the profound ecological challenges facing the planet and growing despair at the inability of traditional politics to address economic failings have fueled an extraordinary amount of experimentation by activists, economists and socially minded business leaders. . . . As the threat of a global climate crisis grows increasingly dire and the nation sinks deeper into an economic slump for which conventional wisdom offers no adequate remedies, more and more Americans are coming to realize that it is time to begin defining, demanding and organizing to build a new-economy movement."[20]

One of the most significant aspects of this growing movement is the challenge it presents to corporate power. Because the driving imperative in the New-Economy Movement is social benefit, not corporate shareholder profit, the possibility of building a balance against the corporate profit-driven politics of the twentieth century is growing.[21] Not since the days of unionized labor pressing for social justice and fairness in distribution of wealth between corporations and workers has there been such a strong voice for the public interest.

Organized labor has been slow to embrace a move to renewable and sustainable initiatives. Strongly grounded in the existing fossil-fueled economy, many unions still resist change. However, some initiatives toward a transition are evident in the Blue/Green Alliance and the Labor Network for Sustainability.[22] While neither of these have yet become mainstream labor voices, the perception that labor has a stake in crafting a sustainable future is growing in some sectors. Unions have yet to define a leadership role in shaping the workforce of the future, though they have the capacity to do so.

Sustainable businesses are also developing an independent voice in politics. The American Sustainable Business Council (ASBC), formed in 2010 as a counterfoil to the American Chamber of Commerce, advocates for policy change at the federal and state level that supports a more sustainable economy. The ASBC spans a growing network of business associations across the United States, which in turn represents over 200,000 businesses and 325,000 business executives, owners, investors, and others. The associated American Sustainable Business Institute, the nonprofit education and research arm of the organization, offers programs that educate and inform the public and policymakers about the benefits of a more sustainable economy, and about policies and practices that can help the economy become more sustainable.[23]

PITTSBURGH PATHWAYS

A new economy locally centered and locally invested springs up in Pittsburgh, and in communities across the country and around the world. Corporations operating for social benefit, not only stockholder profit, expand the distribution of wealth among worker-owners. The well-being of the community is measured in terms of beneficial economic activity, health of the environment, social equity, and cultural diversity. In most cases, the motivation is wholly or in part based on a sense of doing what is right for the future.

ALCOA CORPORATION: A MODEL OF RESOURCE CONSERVATION[24]

Many corporations have pursued a sustainability model as a basis for their business success. Some have designed their operations to incorporate a circular return of raw materials. One early example driven by both the high cost of raw material and visionary leadership sustained over several decades is Alcoa, Inc., one of the largest aluminum processing and manufacturing companies in the world, which has its headquarters and research center in Pittsburgh.[25] Economic parameters have allowed the social responsibility and environmental stewardship aspects of sustainability to align within their corporate profit structure.

Bill O'Rourke, who served in roles from corporate patent counsel to vice-president of sustainability and environment, health and safety during his thirty-six-year Alcoa career, notes that there are over 125 patents on the aluminum can.[26] These changes have increased the ease of use, reduced weight, and engineered the can for recycling. Alcoa, a lightweight metals technology, engineering, and manufacturing company, has a history of buying back aluminum scrap from customers, supporting recycling initiatives, and utilizing recycled aluminum in production. Creating aluminum from virgin sources is an energy-intensive process, and in addition to reducing energy usage and emissions, recycling saves Alcoa money: the cost of reusing aluminum is about 5 percent of the cost of producing aluminum from virgin sources. Because of the very high cost of recovering and refining ore from bauxite, aluminum recycle and reuse reaches about 95 percent in the industry worldwide. O'Rourke said, "Nearly 75 percent of all of the aluminum ever produced since 1888 is still in use today. Increasing efficiency and reducing waste is the right thing to do."[27]

In the late 1990s, leaders took Alcoa beyond recycling and reuse. Patrick Atkins, the first director of environmental affairs at Alcoa, envisioned reducing the company's water discharge to zero, cutting waste, and eliminating mercury emissions. Atkins used his technical expertise to educate others about potential engineering solutions and helped to create Alcoa's 2020 Vision.[28] Former Alcoa CEOs Paul O'Neill and Klaus Kleinfeld also championed environmental and safety goals and created a system for measuring environmental, health, and safety performance. For instance, O'Neill strove to reduce worker injuries through examining inefficiencies and fixing safety concerns in the manufacturing process. As a result, Alcoa not only reduced injuries but also increased efficiency and net income. Today, everyone at Alcoa is responsible for meeting environmental, health, and safety goals, and business objectives and performance reviews reflect these priorities.[29]

These efforts have brought Alcoa to what William McDonough and Michael Braungart call technical metabolism: "a closed-loop system in which valuable, high-tech synthetics and mineral resources—technical nutrients—circulate in a perpetual cycle of production, recovery, and remanufacture."[30] Alcoa was the first aluminum company to receive the Cradle to Cradle Certified designation, a third-party eco-label that assesses material and manufacturing processes and evaluates a product's safety to humans and the environment.[31]

Through increasing the proportion of recycled aluminum in products, Alcoa can continue to reduce the environmental impact of aluminum in products from food packaging to airplane wheels, but one of the greatest challenges to Alcoa's recycling and reuse efforts is the recycling rate in the United States, which hovers around 50 percent. The low level of aluminum recycling in the United States compared to European countries is a clear example of the difference in cultural values. Having a clear vision of the goal of sustainability is missing in the United States. Seeing recycling as an altruistic value rather than an essential part of our economy makes recycling an "externality" rather than a critical part of a dynamic equilibrium. Changing behavior is difficult; however, O'Rourke hopes that efforts of groups like the Aluminum Association and the Alcoa Foundation will educate people throughout the United States about the benefits of aluminum recycling and convince them to act altruistically.[32] Additionally, products need to be engineered for recyclability. Currently, the Alcoa Foundation is sponsoring a free online course about designing products so that they can be returned to industry or nature at the end of their use.[33] Hopefully, these efforts will allow both consumers and designers to rethink waste and to ultimately keep aluminum in a continuous loop of use, collection, and reuse.[34]

ELOOP, LLC: RECLAIMING AND REPURPOSING ELECTRONIC WASTE

One of the fastest-growing sources of waste in the modern economy is tied to the information age and the technology of information systems. It seems that newer, better, faster, prestige versions of electronic devices emerge constantly to make older iterations obsolete. Telephones have gone from fixed installations in buildings connected by wires to mobile devices of all descriptions that perform functions as varied as voice communication and data transfer to entertainment and remote signaling for functions at distant locations. Most of these devices have a useful life of less than two years before becoming waste.

Worldwide, over fifty million tons of electronic waste is generated from discarded computers, mobile phones, television sets, and refrigerators. Only 15 to 20 percent of this reusable material is recycled or recovered each year.[35] In 2012, the United States generated 3.412 million tons of e-waste. Of this amount, only 1 million tons, or 29.2 percent, was recycled, according to the EPA (up from 25 percent in 2011).[36] The rest was trashed—in landfills or incinerators. Much of the electronic garbage ranging from appliances to cell phones and computers ends up in enormous toxic piles in India, China, or other underdeveloped countries where children and unprotected workers pore through the debris to reclaim components for resale. According to the EPA, experts estimate that recycling one million cell phones can recover about 75 pounds of gold, 772 pounds of silver, 33 pounds of palladium, and more than 35,000 pounds of copper.[37] Thus, the issue of recycling and reclaiming electronic equipment becomes an environmental justice issue as well as a resource conservation issue.

With electronics such a mainstream part of the economy and modern life, essential for communications and all levels of information exchange and services from banking to vehicle navigation, the embedded resource use must be considered. A United Nations study found that the manufacturing of a computer and its screen takes at least 530 pounds of fossil fuels, 48 pounds of chemicals, and 1.5 tons of water, more than the weight of a rhinoceros or a car.[38] In contrast with many home appliances, life cycle energy use of a computer is dominated by production (81 percent) as opposed to operation (19 percent).[39] Clearly, the electronics product stream is a system ripe for conversion to a closed-loop process.

Speaking at a Sustainable Pittsburgh Champions 4 Sustainability workshop in 2008, Ned Eldridge, president of eLoop, LLC, discussed the need for a way to recover the materials and resources embedded in electronic waste. Eldridge noted, "Everybody uses electronics today. We have a universal source of product. Our company realizes that corporations and consumers alike have a significant impact on the environment. A sustainable world utilizes the best ways that improve life for many future generations. eLoop's mission is focused around the triple bottom line: environmental, economical, and social benefits."[40] Eldridge's company has a two-tiered approach to recovering electronic waste: First, eLoop will fix, clean, securely remove stored data from, and resell electronics for reuse. Second, eLoop will disaggregate components for reuse.

"One of our unique attributes," Eldridge said, "is our certification of data security for electronic equipment. We practice certified IT Asset Disposition, which allows us to assure data destruction for clients where confidentiality of the data is essential."[41] This allows the devices to be resold without risk of compromising data that may have been stored by a prior owner. eLoop sells reconditioned equipment with a warranty of data security and a profit share of a portion of the resale value back to the client. In a system where planned obsolescence is part of the industry, this restores potentially hundreds of thousands of useful units to the market. According to Eldridge, "When there is a device upgrade, like the constant iterations of the iPhone, thousands of useful devices go out the door. Many consumers just throw them away."[42]

The original equipment manufacturers have to pay for the recycling and recovery of materials, a requirement of a Pennsylvania law passed in 2010.[43] Manufacturers of electronic devices covered under the law must register with the state and pay a fee as well as have an approved recycling plan for collecting used devices and sending them to an approved recycling facility—such as eLoop—that has been certified by the state.[44] The economic impact of paying eLoop for electronic waste disposal is minimal. "Our material processors have the most efficient processes and technologically advanced procedures in the industry. The result is providing an extremely competitive low-cost solution to any organization."[45]

Socially, eLoop understands the importance of recovering electronic waste responsibly.[46] This presents a manpower challenge because of the amount of hazardous components such as lead. Eldridge explained, "We have a very high standard for tracking and managing hazardous materials responsibly. The Basel Action Network standard is an international accord, with a global objective to control hazardous and toxic materials safely. The Institute of Scrap Recyclers, Inc., operates with a weaker standard of responsible recycling, which allows exporting waste to other countries for treatment or disposal. We don't do that."[47] Eldridge said, "We do not send any electronics to developing nations and we focus on keeping the majority of the recycling efforts local. Luckily, this enables us to keep 60 percent of the material in Pennsylvania, creating more green jobs. We are 100 percent transparent, giving consumers and organizations the confidence in knowing the final destination of their electronics. As a Basel Action Network member, we audit all of our electronic waste vendors."[48]

Environmentally, this lowers the carbon footprint with all customers, from individuals to municipalities and schools to manufacturers and businesses. Reusing and recycling all electronics—keeping everything out of the landfill—alleviates the need for mining and starting the manufacturing process from the beginning. Utilizing existing material continues the loop, which uses less energy.[49]

Eldridge explains that the Pennsylvania enabling legislation adopted in 2010 with the help of PennFuture provides for the recycling of a list of "covered devices," which is pretty short compared to California's, but better than nothing.[50]

This approach to electronic waste recovery has several impediments, however. Eldridge explained, "Our biggest problem right now is that scrap prices are at a twenty-year low. The original equipment manufacturers drive the prices down as much as possible to control their own costs, and the Pennsylvania law sets floors rather than maximum targets for recycling electronic waste."[51] Designing the electronics industry around the goal of recovering and reusing materials would solidify the back end of the cycle. "We need a stable market for the recovery products for this to work at its full potential," said Eldridge. "Until then, we cut costs as well as we can, and market broadly to capture as many resources as we can."[52]

THE UJAMAA COLLECTIVE: THE SPIRIT OF COMMUNITY RISING

Sometimes inspiration and action rise up from what seems to be the pit of despair. Women caring about each other, about their community, and wanting an outlet to make things better can change the world. It has happened again and again, from the women's suffrage movement to Wangari Maathai's Green Belt Movement. The Ujamaa Collective started with a prayer circle of a group of women trying to encourage each other, to help each other, and to keep the courage to move forward alive. They began a microinvestment process, passing the hat in their meetings, with one person each time in rotation receiving the proceeds to use as she might need. The concept was based on a spirit of cooperation for the benefit of the group as a whole.

In February 2008 Celeta Hickman took the spark of inspiration that was to be the Ujamaa Collective and put forth a call to action.[53] The idea was a simple one: black women, entrepreneurs, acting in unity for the benefit of the individual entrepreneur, the collective, and the community. Integral to this concept was the inclusion of a green marketplace that would benefit Pittsburgh's Hill District community physically, economically, socially, and spiritually. At first, the response was slow, and meetings consisted of only a small network of sisters who felt the urgent need for the idea Ms. Hickman proposed. However, Ujamaa soon experienced an exponential growth that expanded its intent into a multifaceted collective idea, with input from many voices and visions. Since spring 2008, the women of Ujamaa Collective have hosted, partnered, and supported over a hundred programs and exhibits in surrounding communities of Pittsburgh.[54]

The Hill District of Pittsburgh is an area that once thrived in the 1920s with a rich history of African American culture in music, writing, and the arts. Urban development in the 1960s with major highways diverting traffic away from the area, and cutting off access from the center of the city, led to a period of decline, disinvestment, and population loss. Now the picture is turning around. The library at the intersection of Centre and Kirkpatrick Street was designed based upon neighborhood requests—a sunny reading room, plenty of computers, and a strong section of African American literature—and has been a resounding success. Its opening in 2008 also kicked off a wave of development that hasn't stopped. Fifth Third Bank opened a branch on Centre Avenue in 2009, providing another banking choice for the area. The Hill House reopened a renovated Kaufmann Center, with the stunning Elsie H. Hillman Auditorium, in 2011. The YMCA opened a sparkling new $13 million facility in 2012, Shop 'n Save opened a grocery store in 2013, and Dollar Bank opened in 2014.[55]

The Ujamaa Collective manages a fifteen-acre urban farm in the Hill District, and has launched an Entrepreneurial Preparation Program geared toward the unique needs of women of African descent within the Pittsburgh community, with a concentration on serving the Hill District. LaKeisha Wolf, executive director of Ujamaa, explains why it is so important for people to see that their efforts can create good results that have a direct benefit in the place where they live. "Becoming self-sufficient in the most basic things like feeding your own family from your own community garden is a transformative experience," she said.[56]

The groundbreaking for the Ujamaa Collective Marketplace and opening of the Ujamaa Boutique happened in 2010, offering the high-quality and handmade goods of the women of the Ujamaa Collective as well as other local and international artisans to the Pittsburgh region. Ujamaa Collective's Open-Air Marketplace is poised to become a regional destination for visitors and locals alike. From April through October, the family-oriented, cultural events bring together over 1,500 community residents to share organic produce, handmade crafts, various arts, education, music, dance, and healthy food choices.[57] It offers a space and a connection point for people within the community and attracts people from outside the community to share the experience of shopping, music, dance, stories, and conversation. People can get to know each other in ways that are not possible in an impersonal box store.

The Ujamaa Marketplace creates a regional model of environmental justice. The work of the Ujamaa Collective created a thriving community of people connected through culture, food, and enterprise where there had been despair, blight, and neglect. The spark of determination fueled by hope and encouraged by its own success has transformed and empowered this community. It offers a colorful open-air shopping destination where abandoned lots once stood, providing quality products created locally and primarily supporting the efforts of African American women artists, producers, and manufacturers. The Ujamaa Boutique, a women's cooperative enterprise established as a year-round indoor shop on Centre Avenue, now thrives as it nurtures entrepreneurs and artisans, expanding the connection to the entire city and, through internet connections and personal visits, to the world.[58]

The vibrant spirit threaded with music, stories, and a passion for connecting people with the earth and with each other marks this group as an inspiring example of success. Several Ujamaa Collective members are in ABAFASI, a "wimmin's drum circle" in the West African tradition, founded by Kelly e. Parker.[59] They are frequent performers in the Ujamaa Marketplace and all around Pittsburgh. When they sang about "Gran'ma's Hands," at the Freedom from Fracking concert, the sentiment of caring connected across generations, across the barriers of race and class, building bridges to the future.[60] What was once an abandoned lot rings with the sounds of song, laughter, conversation, and commerce that is fair trade, healthy, handmade, homemade. The Ujamaa Collective gives testament to the value of a diverse and inclusive community acting through "Cooperation over Competition."[61]

CONCLUSION

> *Sustainable development will require a change of heart, a renewal of the mind, and a healthy dose of repentance. These are all religious terms, and that is no coincidence, because a change in the fundamental principles we live by is a change so deep that it is essentially religious whether we call it that or not.*
>
> **HERMAN E. DALY**

We stand in a moment both terrifying in the possibility of crashing life as we have known it and exhilarating in the potential for new human achievement. A new future lies in waiting, wrought not of the fires of competition over the earth's buried treasures but of the effervescence of cooperation as creatures sharing a journey through space and time on a wonderfully complex, living planet. The choice lies between structuring civilization within the constraints of earth's living ecosystems or continuing on the course that leads humanity ever closer to extinction.

Mayor Bill Peduto speaks at a Climate Action Now meeting in Pittsburgh.
MARK DIXON, BLUE LENS

9
BUILDING A RESILIENT CITY

Pittsburgh and other cities are on the front lines of the climate change crisis, and it is our responsibility to address the deep challenges it is creating for us, our children and our grandchildren.

MAYOR BILL PEDUTO

THE TRANSITION TO sustainable practices entails simultaneous solutions to social justice and equity issues, environmental issues, and economic issues. The current situation imposes a series of inequities and injustices based on the imbalance in the sources and uses of the earth's resources. Transition to an economy structured around renewable energy resources, regenerative agriculture, and circular green-chemistry-based materials management will require a shift in both economic and government policies. Calls for social equity and environmental justice mark this time of transition, often against harsh realities. The twenty-first century brings a convergence of rapid global communication, technology for harnessing distributed energy resources, and the recognition that continued combustion of fossil resources cannot be sustained.[1] Both climate change and global pollution are on a time-sensitive path, but resolving the layers of inequities inherent in the current system requires transformative kinds of thinking.

Just as the spread of telecommunications into developing countries did not follow the same infrastructure path as the hard-wired telephone system in the United States, so the transformation to a renewable and distributed energy system can progress without all the intermediary steps. As a national priority, America can help emerging countries skip the fossil stage and move right to the renewable system, as telecom went straight to wireless and decentralized systems. Why should a community in rural Kenya or Nepal build out a rural electricity system based on fossil-fuel-powered electric plants connected by wires over long distances when they can install microgrid solar systems with on-site batteries? In China, for example, the combination of government investment policies, health effects from fossil fuel pollution, and increasing demand for energy in rural areas has launched global renewable energy leadership. Analysis of the situation by Stratfor, a geopolitical intelligence firm, finds that "China has already passed the tipping point where renewables are coming online faster than demand is growing, and in 2015 consumption of fossil fuels fell by 1.4 percent. In the same year China accounted for 28 percent of global electric vehicle sales, 32 percent of solar panel installations and 47 percent of wind installations; in 2016 it overtook Europe in wind power and by 2020 will probably surpass it in solar power, too."[2]

Technical solutions for sustainability alone are insufficient to create change without political and social institutions aligned to expedite the transition. The technology is available for buildings that operate at net-zero energy, water, and waste, but multiple hurdles in the US regulatory system confront each project. Most of the existing framework of building codes, zoning ordinances, and financing standards was not designed for these kind of integrated systems. The problem is especially difficult when the goal is to achieve such standards for a business development or a community project where funding is a constraint. Effective implementation of sustainable energy, agriculture, and materials production will require a realignment of government policies.

Major changes rarely take place without supportive government policy in response to massive underlying public and societal support. Sweeping reform of existing policies is unlikely without a transformative vision to inspire a broad consensus on action. In the absence of such a broad national vision, adaptation to changes that are now inevitable and programs for resilience in the face of change are emerging from communities that share concern for each other and for the future. The regulatory infrastructure to ease this process into mainstream practice emerges one community at a time. Revisions in regulations, codes, and ordinances will follow in response to new visions of what communities want and need for environmental, social, and economic sustainability; resilience; and long-term viability. Ethical choices drive the policy changes and build new pathways forward.

Pittsburgh has taken up a path toward a more sustainable and resilient community by both aligning community initiatives and addressing the infrastructure to support change. In this effort, the need for equitable solutions plays a prominent role. Some initiatives come from long-standing traditional industries, and some emerge through entrepreneurial efforts in both the nonprofit and business arenas. A full discussion of such initiatives would take many volumes, as a snapshot of the city can show.[3] But such transformations rarely occur without some unifying governmental leadership and action, and the key component of this initiative for building a resilient and sustainable city comes from the leadership of a visionary mayor. Pittsburgh mayor Bill Peduto made his vision clear when he said, "What can we do to do more together? Look at not 'What is in this for me,' but 'What is our collective potential?' We need to plan for facing the uncertainties the twenty-first century will bring. We need to plan for upgrading the infrastructure in ways that make the environment better, not worse. We need to assure that everyone who lives here has the opportunity to participate in the new economy."[4]

THE CHALLENGES TO A SUSTAINABLE FUTURE

Few major changes in the energy arena take place without government impetus. Whether at the initial stages, as with the interstate highway program that opened transportation and enabled the suburbs, or after initial market development revealed public necessities, as with the Federal Power Act, government action can accelerate and organize change. This is a time of opportunity for government to act to streamline the transition away from fossil fuel dependence, and to address issues of equity and fairness.

Pittsburgh has already taken bold steps to lead on climate change. Mayor Bill Peduto participated in the mayors' summit in advance of the COP21 talks in Paris. At the summit, Peduto presented the city's next steps—by the year 2030—for reducing its emissions through renewable energy consumption, energy and water conservation, fleet conversion, landfill diversion, and transportation emission reductions. Those goals—covering city government-owned facilities, fleet, and infrastructure—for 2030 include:

- 100% renewable energy supply
- 50% energy consumption reduction
- 50% water use reduction
- 100% fossil-fuel-free fleet
- 100% diversion from landfill
- 50% citywide transportation greenhouse gas emissions reduction
- Creating a fossil-fuel-divestment strategy for City of Pittsburgh funds[5]

This is in addition to the Pittsburgh Green Building Alliance 2030 District plan, in which the city, Allegheny County, and related authorities have been engaged for the past three years. Through the Green Building Alliance 2030 District process, more than 70 percent of properties in Downtown and Oakland—both publicly and privately owned—are voluntarily reporting their energy use, more than thirty different property types are reporting their energy or water use, and public and private partners are supporting studies of transportation levels and emissions.[6]

Peduto has made sustainability and resilience a major priority for the city since his inauguration. His enthusiasm and persistence in the face of the underlying press toward Marcellus Shale gas development has inspired entire communities within the city to take new initiatives around renewable energy, urban agriculture, and healthy, safe communities. Each year more people are engaged, more businesses step forward, more institutions and neighboring municipalities come forward. Nevertheless, major issues remain.

In Pennsylvania, mainstream policy heavily favors fossil-fuel-based options, especially since the Marcellus Shale boom seems well aligned with the skills and infrastructure of society from past days of perceived greatness.[7] As of 2015, the Allegheny Conference on Community Development summarized their view of the area's energy future primarily as a development opportunity: "The Pittsburgh region has emerged as the new Center of American Energy in the United States. The region combines historical strength in energy resources with emerging strength in innovation and energy solutions. We are home to the nation's first oil and gas wells, its first commercial nuclear reactor, a rich coal seam and now the expansive Marcellus and Utica shale natural gas plays."[8] Although The Allegheny Conference planning does include energy-efficient buildings among the "innovation" initiatives, the main thrust of the business focus remains on the Marcellus Shale gas and "clean coal" technologies.

New opportunities for chemical production from cheap, abundant methane sourced from the Marcellus and Utica Shale formations raise hopes and expectations of a return to times of prosperity for people who once worked in the steel mills. But voices of environmentalists and health advocates urging that the opportunity be taken to prevent the smog and air pollution of earlier times clamor for precautions. Shell Chemicals plans for a refinery to manufacture polyethelene plastic precursor materials using Marcellus Shale gas as a feedstock in Beaver County feeds the vision of jobs returning to areas abandoned by manufacturers in the last decade. The refinery operations will increase the air pollution because the new plant bought air pollution credits from operations that have closed in the last ten years.[9] The plant air emissions include volatile organic compounds including benzene, a known human carcinogen. The products of the plant will add to the plastic throwaway society with no design for reuse or reclamation of the material drawn from the fossil source.

However, the prospect for new jobs in areas suffering decades of blight shunts aside those who call for protection for the environment and for the communities in proximity to the proposed refinery. The environmental justice concerns around existing coke works, coal-fired power plants, and chemical production facilities continue to present challenges. While some activists and environmental advocates work to clean up the pollution, others focus on the needs of workers and the rights of businesses and property owners. But unlike the times of the past when workers had little say in the conditions of their employment, and communities did not speak out for the living conditions of their citizens, modern communities such as Pittsburgh have fought hard for environmental improvements and better conditions for living and working.

The shift in the paradigm for utility services is happening, and it is most perceptible at the interface of customer efforts to increase efficiency and reliability. Increasingly, customers have choices now that include some measure of producing their own energy by becoming as efficient as possible, and by adding solar, wind, or geothermal options, directly or by choosing a renewable electricity supplier. However, many current utilities see customer-based distributed generation as a threat, and many in the industry talk about the "death spiral" of customers becoming their own power providers.[10] Overcoming this barrier will require changes in law.

The struggle between jobs, regardless of the source, and environmental quality continues as a fierce and sometimes polarizing conflict. This ethical problem of value for the climate and living conditions of future generations versus the immediate needs of workers and families for jobs presented by Marcellus Shale gas development and associated infrastructure such as refineries, pipelines, and compression and export facilities divides communities and even families. With a national "America first" energy strategy with reduced regulatory constraints for environmental protection, these conflicts will continue.

LAYING THE GROUNDWORK FOR CHANGE

Leadership is critical to the success of city transformations. Mayors in cities like Pittsburgh are in close proximity to the problems and to the people, making the cause and effect connections easier to visualize. Pittsburgh is fortunate to have Mayor Peduto leading this quiet revolution. His transition team invited city residents to come together and share their visions for their own neighborhoods, and 1,400 people participated directly. The mayor held workshops around critical issues like public safety, education, water management, accessibility of government, and healthcare. Working groups and task forces began developing the details. Public-private partnerships were formed with the business community, with the universities, and with civic organizations. The philanthropic foundations of Pittsburgh, many founded from the wealth of the industrial age in steel, glass, coal, and finance, provided seed funds and resources outside of municipal resources. All around the city, people sat down together to work on problems and craft creative and exciting solutions.

Drilling down to the sources of the problems of energy, water, waste, and food supply offers the opportunity for new solutions based on caring for the future. The international 100 Resilient Cities initiative of the Rockefeller Foundation is one program that encourages such an approach. The Pittsburgh Plan for Resilience, as part of the 100 Resilient Cities initiative, sets out the challenge thus:

> Pittsburgh faces fewer threats than many other cities thanks to our location, geography, and natural resources. However, the Steel City must still overcome certain challenges from its industrial legacy, and will face new pressures with climate change, urbanization and globalization. Pittsburgh will be a resilient city when our entire community shares in the same opportunity and prosperity, and all residents are equally well cared for and well prepared to face potential risks. Urban resilience is the capacity of individuals, communities, institutions and businesses within a city to survive, adapt and grow no matter what kinds of chronic stresses and acute shocks they experience.[11]

Pittsburgh is typical of many midsized cities that experienced a traumatic shift from a center for manufacturing and industrial production to a community with an economic base founded on services. Between 1950 and 1990, the city suffered population losses and deteriorating infrastructure with abandoned buildings falling to ownership by the city by default. After World War II, personal car ownership and suburban life in single-family residences exploded the population outside of the city. Cities were made friendly to vehicles, and people moved farther out of the center. In some areas, adding express highways to expedite the growing number of personal vehicles on the road had the effect of chopping up and isolating or bypassing city neighborhoods. These suffered from disinvestment and decline as the neighborhood economic base eroded.

An economy based on a dynamic equilibrium would align the use of space to colocate living, working, and recreational uses. Reclaiming and reconfiguring existing cities can preserve and renew the elements of people-centered cities. Many younger people want to live in a place where their residence, work, recreation, and conveniences for food and amenities are close together. They want flexible transportation options. Their sense of identity is not bound up in a car as a status symbol.[12] The model of a city responsive to the needs of the twenty-first century includes a new look at land use, recreation and cultural amenities, and achieving diversity and equity in serving the needs of people as part of a community.

What gives hope and inspiration to the messy process of evolving to a new state of economic equilibrium is the culture of communities of caring emerging from so many parts of the city. An environment of encouraging exploration and empowering people to engage in developing their own solutions bears fruit in the form of innovation. Pittsburgh offers a model for the process of recreating a resilient city for the twenty-first century.

LARIMER: A COMMUNITY VISION IN ACTION

The Pittsburgh neighborhood of Larimer was named for William Larimer, whose manor house overlooked what is now Larimer Avenue and who used rents from property in this area to finance coal operations in Westmoreland County. Originally settled by Germans in the 1800s, from the 1900s to the 1960s the neighborhood was known as "Little Italy." Then urban redevelopment road improvements and a large housing project of the early 1970s, plus the flight of people to the suburbs, isolated this community and led to its decline. No vestiges of the former Italian community remain in the area now.[13] After thirty years of decline and multiple past planning initiatives, the citizens and partners of Larimer began organizing in 1998 to create projects that would change the community from the inside out.

Through a decade-long process supported by community leaders and businesses, and with help from Senator Jim Ferlo and the Pittsburgh Urban Redevelopment Authority, a community action plan emerged. In the process of developing the plan, people in the community learned to work together, build trust, create immediate results, and position community members as leaders in the future development of the neighborhood—all of which helped to attract the resources necessary for a strong, vibrant, and safe place to work, worship, and raise families. The first priority for year one of the community action plan in 2008 was "Building trust and capacity from within."[14]

Driven by The Larimer Consensus Group (LCG), a gathering of 125 stakeholders including residents, business owners, community organizations, and partners, the community continued its efforts and developed a strategic plan, the Larimer Vision Plan.[15] This comprehensive transformation plan addresses three key aspects of a neighborhood: housing, people, and the community. Its purpose is to synthesize multiple strategies that work better because they work together. Work on the Larimer Vision Plan brought stakeholders from the Larimer neighborhood together in order to be at the forefront of community-driven change. The process gave voice to people who had not been involved in plans for the community in the past.

The Larimer Vision Plan steps back from the details of individual streets and buildings and looks at land-use pattern and potential. The Kingsley Association, a community service organization dedicated to rejuvenating the Larimer community, has taken a leadership role in this plan. "We are one of the best flat places in this city of hills. We look for the best use of the land, since so much of what is here is vacant or derelict, it is almost like a clean slate to plan for the best sustainable solution," said Fred Brown, the associate director for program development for the Kingsley Association in Larimer.[16] A major goal is to establish a new identity as a state-of-the-art green community.[17] "We have a vision and a goal and the means to start on this ambitious process, but much of what we want to do will need about twenty-two zoning variances," Brown noted.[18] The solution involved collaborating with the City of Pittsburgh to make a new zoning process.

Larimer received one of four national $30 million Choice Neighborhoods Implementation grants from the Department of Housing and Urban Development in June 2014.[19] The grant will help construct housing to advance the neighborhood's plan for a green, sustainable future. Brown described the process of moving this community forward: "It has been a sustained effort, and we have had to overcome many obstacles." He noted, "If there is ever a situation where you want to put up buildings that are as energy-efficient as possible to control operating costs for the occupants, it is when you are building affordable and low-income housing. However, the standards and conditions associated with federal grants make solving for multiple efficient systems very challenging."

Many of the challenges Larimer faces—such as high vacancy, an older housing stock, and a shrinking, aging population—are problems faced by many communities across the Pittsburgh region and across the country. The opportunity here is to address the very real problems of the neighborhood in a manner that is a state-of-the-art model for the Pittsburgh region. Brown said, "We are in a good position to have housing, recreation, business, and urban agriculture in ways that are green and sustainable. Because we are close and connected already with transportation corridors to parts of the city that are bursting with businesses and opportunities, Larimer is ready to become a model for a twenty-first century, resilient, sustainable community."[20]

UPTOWN ECODISTRICT: TRANSPORTATION AND ENERGY SELF-RELIANCE

Pittsburgh has been accepted as part of an internationally recognized EcoDistrict program that connects issues of land use, energy, and food systems in communities.[21] City of Pittsburgh sustainability and resilience officer Grant Ervin said, "The city wants to initiate the EcoDistrict concept in the Uptown neighborhood, located between Oakland and downtown to improve the safety of biking and walking in that community."[22] Uptown sits at the center of two of the most dynamic real estate markets in Pennsylvania. Downtown Pittsburgh and Oakland are the second- and third-largest commercial centers behind Center City, Philadelphia. The area is also home to two large universities, a major medical center, and a professional sports venue. However, like many neighborhoods in Pittsburgh, Uptown has seen levels of disinvestment and deterioration that are not befitting such a well-connected neighborhood. The population has dwindled to just over eight hundred permanent residents.

The resulting empty lots and buildings have created severe losses in property values. Because of the amount of abandoned lots, the most profitable land use has become surface parking. Less street activity has further fostered a perception of an unsafe environment. Though industrial uses provide jobs, they have many blank walls and chain-link fences that don't create a welcoming urban experience. Although many transportation options exist, the street network isn't pedestrian- or bicycle-friendly. Fast-moving traffic and unbuffered sidewalks make walking and biking unpleasant and dangerous.[23]

Through the City of Pittsburgh Sustainability initiative, a team of local civic and government leaders began to examine ways to update and streamline Bus Rapid Transit through this area while taking the abandoned and vacant lots surrounding the corridor as opportunities to recreate a viable neighborhood. "We wanted to build a new Uptown that is sustainable, economically viable, and ecologically sensitive," said Ervin.[24] A team was formed that included representatives from the City of Pittsburgh, Carnegie Mellon University Remaking Cities Institute, Sustainable Pittsburgh, Uptown Partners of Pittsburgh, Oakland Planning and Development Corporation, and the Urban Redevelopment Authority of Pittsburgh's Center for Innovation and Entrepreneurship, and community interests.[25] In addition to traditional economic development indicators, the Uptown team's goals included economic and equity issues, such as opportunities for wealth building, affordability and economic diversity, and ecological aspirations for improved air quality, resource efficiency, and green space connectivity. The entire team attended an EcoDistrict Incubator as part of the Class of 2014.[26] "Our main reasons for attending were to learn about the EcoDistricts process and explore how we could apply the process to the Uptown corridor," said Ervin. "It was also a team-building exercise to get members of the team all on the same page." After less than a year of working together, according to Ervin, "we are rocking and rolling!"[27]

As the community team worked together, a shared vision with twelve specific initiative areas emerged. Four specific projects were chosen as catalytic initiatives to help coalesce the vision and demonstrate visible evidence of action in the wider community. There are now three stakeholder groups working on the Bus Rapid Transit (BRT) concept, the district stakeholders, and a technical advisory group. The Uptown Partnership formed a formal memorandum of understanding between the City of Pittsburgh, the Port Authority of Allegheny County, and the Redevelopment Authorities of Pittsburgh and Allegheny County to work collaboratively on the BRT project. The BRT project would reduce the traffic congestion along one of the most heavily traveled corridors in Pittsburgh while improving access and connections among the major business and cultural elements. Redirecting travel to electric buses will not only reduce congestion, it will improve air quality from vehicle emissions and make pedestrian and bicycle travel easier.[28] "Now, we are looking to share what we learned with the community, and set up a framework for other potential districts," said Ervin.[29]

Peduto sees the Uptown project as the first in a series of efforts to rebrand areas of the city that have fallen into decline as opportunity zones. "I know the economic centers of Oakland, Shadyside, and Downtown will be fine on their own with plenty of investment from the private sector moving opportunities forward. I want to see all of the city of Pittsburgh become a vibrant, resilient, ecologically healthy, and safe place. I want to see the EcoDistrict concept move to a net zero energy, water, and waste concept from the Monongahela River banks to East Liberty and Larimer—off the grid, if possible!"[30] He sees that

Pittsburgh has all of the elements for success to become a model city for the twenty-first century. Peduto explained:

> We can work one community at a time into an energy system of district heating and cogeneration supplied by renewable resources including biofuel generated in the wastewater treatment facility. With the intellectual capital of all our universities, we have access to the most advanced ideas and technologies in the world. I have entered into a Memorandum of Understanding with Carnegie Mellon University to provide dedicated research modeled in Pittsburgh that we could share with other cities and with the world. We can create and implement best practices for people, the planet, profits, and place.[31]

Peduto noted in particular that Pittsburgh has always been a special place with a history of leading innovation in practice. "We are building a future that is neighborhood based and people friendly, not focused around cars, but around the needs of communities. It depends on involving lots of people for this to work."[32]

Peduto sees the next stage as putting the regulatory and institutional infrastructure together:

> We will need to update or overlay a zoning code to support the EcoDistrict concept. People should not have to go through endless variances to implement innovations. We will also need to see about streamlining the utility tariffs and go to the Public Utility Commission if necessary to have an integrated energy system that serves the needs of the future. The key to all of this is to keep people involved. I hope to establish a volunteer commitment as part of the culture of this city. We can challenge the entrenched inertia and empower people to take part in solving problems together. Unleash the pent-up desire for action through partnerships. There is a place for everyone to be part of this.[33]

Developing a new base for aligning the zoning and building codes to make new system solutions possible is underway. It will be a process that requires broader thinking, and adjustments in traditional ways of doing business. But the new dynamic in solving multiple problems simultaneously, rather than individually and separately, can open broad avenues of innovation. By formally reaching out to tap the intellectual capital of Pittsburgh's universities and research institutions, Pittsburgh becomes an urban laboratory for designing the social infrastructure to support a leap forward into a sustainable future.

PHIPPS CONSERVATORY AND BOTANICAL GARDEN: A LIVING BUILDING

Imagine a building designed and constructed to function as elegantly and efficiently as a flower.

LIVING BUILDING CHALLENGE

The history of this area is replete with the evidence of the fossil age, up to the recent press for fossil gas extracted from the deep shale of the Marcellus and Utica formations that underlie most of western Pennsylvania on up into New York. But Pittsburgh also boasts the first air pollution controls, introduced in 1946; is home to the first net-zero-energy public building; and stands eighth in the nation in Leadership in Energy and Environmental Design (LEED)–certified buildings. There are currently thirty-nine LEED-certified buildings in the city and more than fifty-nine in Western Pennsylvania, with over sixty new projects in the city pursuing LEED certification.[34]

As the struggle toward a more just and equitable future continues, system solutions that address multiple problems emerge. Community spaces can model change by allowing many people to experience an energy system designed to conserve energy, water, and waste. The power of experiencing a new energy system can help people question the status quo, and, framed in a context of accountability and responsibility to the future, such community projects can stimulate broader acceptance and support at the policy level.

A building that generates its own energy might sound like science fiction. However, unlike flying cars or time machines, buildings like Phipps Conservatory and Botanical Garden's Center for Sustainable Landscapes are designed to use the sun to create their own energy. This means that the building has net-zero energy impact or is operating with energy from the sun, wind, or earth to meet or exceed net annual demand. How the Center for Sustainable Landscapes building came to generate more energy than it uses is due to a box, a bold idea, and an integrated approach.

Henry Phipps, partner in the Carnegie Steel Company, built the Phipps Conservatory as a gift to the City of Pittsburgh in 1863. It operated as a place of public beauty, open on Sundays for the benefit of workers. Under management as a city asset, Phipps was a place of interest for generations, but in the late 1980s it fell into disrepair and decline as budget problems struck with the collapse of the steel industry. Finally, the City of Pittsburgh turned the facility over to a newly formed nonprofit organization in 1993 with a one-hundred-year lease. Richard V. Piacentini came to Phipps as executive director in 1994, bringing a fresh perspective and great dreams. He has been instrumental in improving the conservatory's collections, educational programs, and overall services to the public. During his tenure, Phipps launched the most ambitious expansion project in its 113-year history.

Piacentini began reimagining the Phipps Conservatory with a new Phipps's visitor center project, which he intended to be the first LEED-certified visitor center. During the initial stages of this green construction project, Piacentini stopped and noticed a box of tiles for the new visitor center. The tiles had been made in another country. When Piacentini asked why the team had purchased tiles that involved significant transportation, he was told that the project already had earned enough LEED points in the materials supply category. In that moment Piacentini realized that the LEED Certification process wasn't just about the points. Instead, Piacentini shared his ethical vision for the project: "Phipps is doing this because we think this is the right thing to do, and from that moment on we started to look at everything we did. We tried to be as efficient and green as possible. Phipps can be transformed to a model of sustainability for the next century."[35] Phipps planned this LEED-certified building expansion to reflect their values of energy and water efficiency, sustainable development, and high operating performance.

Piacentini learned about the Living Building Challenge at the Greenbuild Conference in November of 2006, where he met Jason McLennan, the founder of the concept.[36] The Living Building Challenge, now a project of the International Living Future Institute, visualizes designing a building to function as elegantly and efficiently as a flower. The concept has seven "petals" that define performance areas: place, water, energy, health and happiness, materials, equity, and beauty.[37] Piacentini thought that this idea would be especially well suited for the Phipps campus,[38] as the mission of Phipps is "to inspire and educate all with the beauty and importance of plants; to advance sustainability and promote human and environmental well-being through action and research; and to celebrate its historic glasshouse."[39] A building that acts like a flower relying on current solar income for energy would certainly advance sustainability and promote human health.[40] Piacentini knew that funding this kind of innovative project would challenge his board. He proposed a three-phase process to build a model for sustainability: The first phase reimagined the visitor welcome center, the second added the Tropical Forest Conservatory, and the third built the Center for Sustainable landscapes.[41] The new Visitor and Welcome Center opened in 2005, and the second-phase Tropical Forest Conservatory opened in 2006.

After drawing his napkin sketch of the living building, Piacentini shared what he learned about the Living Building Challenge with the Phipps board, intending to progress to phase three, the Center for Sustainable Landscapes, gradually, out of concern for fundraising.[42] However, after seeing the second phase of the Phipps plan come to fruition in December 2006, donors wanted to know what was to come next, and expressed their interest in supporting the next transformative step in the Living Building Challenge.[43] By remaining committed to the sustainability goals and continuously raising the bar on green building standards, Piacentini had enhanced the donors' desire to invest in an organization that was on the cutting edge of sustainability. Even with the economic crisis of 2008–2009, Phipps was able to raise enough funds for the building in 2010,[44] achieved Net Zero Building Certification for the Center for Sustainable Landscapes in February 2014, and in 2016 is pursuing Living Building Challenge certification.[45]

The process for building the Center for Sustainable Landscapes involved collaboration with experts including architects, engineers, contractors, and students from two local universities. There were fifteen design workshops among the architects, designers, and stakeholders. In fact, as in an ecosystem, each expert had a unique role to play in the process, and each system from water to energy was interdependent and integrated. Therefore, each member of the team needed to communicate needs and constraints to create this one-of-a-kind building system that creates its own energy, cleans its own water, and enhances the lives of occupants and visitors.[46] Vivian Loftness from the Center for Building Performance and Diagnostics at Carnegie Mellon University noted, "Without an integrated process it's not possible to build something that is cutting edge and certainly not something that is trying to set a standard for the future like the Center for Sustainable Landscapes."[47]

In reflecting on the process, Piacentini noted that this Living Building approach can work with all kinds of building projects. "People make cost and value decisions all the time in building projects. It is just a matter of deciding what is most important: a marble lobby or a zero net energy profile? It is not about money; it is about what we value."[48] Piacentini noted, "Visitors don't come here just to see pretty flowers anymore. They come to see what a building designed in harmony with nature looks like. Anyone who comes through the Center for Sustainable Landscapes would love to work in a building like this."[49] Phipps Conservatory and Botanic Gardens and Center for Sustainable Landscapes stands as the true "green heart of Pittsburgh."[50]

The value of the Phipps project extends far beyond this single site. Phipps's success has inspired and motivated many other projects in Pittsburgh including the zero-net-energy campus of Chatham University at Eden Hall, the Frick Environmental Center in Frick Park, and the new Forest Hills Borough building.

CONCLUSION

> *Mankind has gone very far into an artificial world of his own creation. He has sought to insulate himself, with steel and concrete, from the realities of earth and water. Perhaps he is intoxicated with his own power, as he goes farther and farther into experiments for the destruction of himself and his world. For this unhappy trend there is no single remedy, no panacea. But I believe that the more clearly we can focus our attention on the wonders and realities of the universe about us, the less taste we shall have for destruction.*
>
> RACHEL CARSON

The cumulative effects of human enterprise since the Industrial Revolution have caused an upheaval in the global conditions that provided a stable environment for life to thrive for the last ten thousand years. Changes wrought by the reaction of natural systems through the universal laws of physics, chemistry, and biology spin global conditions into uncharted areas. In the face of certain but unknown change, nations respond either with vision and forward momentum to address the issues of climate change and pollution, as China has, or they retrench into familiar patterns and resist adaptation, as the United States is doing under President Trump. Where visionary leaders step forward to empower people and to recognize the importance and value of preserving the natural systems of the living earth, great possibilities emerge. In the absence of national leadership, the initiative falls to states and cities across the United States as people clamor for action.

Pittsburgh is such a place, designated a Biophilic City in September 2016 for its commitment to "eliminate the use of all pesticides, fungicides and herbicides, to increase the city's tree canopy from 42 percent to 60 percent by 2030, to pursue when possible the daylighting of streams in stormwater management efforts and to develop more greenways."[51] In accepting the designation, Mayor Peduto said, "Our commitment to clean water is not just to build bigger pipes but to restore streams to our valleys. . . . We will no longer use pesticides, so our workers will be weeding by hand. We expected pushback but our workers were carrying 40-gallon jugs of poison on their backs. There has been no pushback, because they understand."[52] This designation symbolizes the recognition of the people of Pittsburgh that a healthy environment is essential for a thriving community.

E. O. Wilson argues that to preserve life on Earth fully half of the natural systems must be preserved in undeveloped state to assure the continuation of essential ecosystem services.[53] As the reality of climate change and the health effects of pollution become increasingly difficult to deny, people seek leadership to choose a different way forward. In Pittsburgh, Mayor Peduto's approach empowers many people by engaging them to develop solutions one neighborhood at a time. When people see the changes they have helped to create, they are committed to do even more.

A transition to a sustainable future entails many incremental changes, and some large systemic shifts. These will not happen overnight. But in every instance where a community, business, organization, or institution makes a commitment to sustainability, the transition advances. Pittsburgh carries on its tradition as a leader in innovation spurred by active partnerships between the city and the universities, and by bold businesses and institutions that carve a new way forward even in the face of tenacity from older industries. Pittsburgh, once the smoky engine of the Industrial Revolution, is now a Biophilic City, blazing a clean, resilient, inclusive pathway into the twenty-first century.

CONCLUSION

THE POWER OF JOINED VOICES

Mitigating climate change and adapting to its effects are necessary to eradicate extreme poverty, reduce inequality and secure equitable, sustainable economic development. . . . Climate change is intrinsically linked to public health, food and water security, migration, peace, and security. It is a moral issue. It is an issue of social justice, human rights, and fundamental ethics. We have a profound responsibility to protect the fragile web of life on this Earth, and to this generation and those that will follow.

BAN KI-MOON

It took hundreds of millions of years to produce the life that now inhabits the earth—eons of time in which that developing and evolving and diversifying life reached a state of adjustment and balance with its surroundings. Given time—time not in years but in millennia—life adjusts, and a balance has been reached. For time is the essential ingredient; but in the modern world there is no time.

RACHEL CARSON

THE TWENTY-FIRST CENTURY is well underway. By its end, the conditions of earth will be vastly different, in part from the progression of the planet in its orbital trajectory around the sun but also from the collective actions civilization will take. The existential crisis presented by climate change and global pollution requires a mobilization effort equivalent to waging a war. There is already much war in the language surrounding these issues: "the war on coal," "water wars," "war on hunger," "war on terror." In fact, it is the war on nature that has proceeded without restraint for decades. This war is waged as slow violence, cut off from view, ignored, and sometimes even denied in the name of progress and economic growth. Just as moral outrage over the atrocities of war coalesce human response to quell tyranny, the ethical imperative to prevent the deliberate destruction of the earth has arisen on behalf of the children of the twenty-first century.

Time is of the essence. Every atom of fossil carbon released from its bound state in the earth's crust becomes a hundred-year burden on the atmosphere and ocean carbon reservoirs of the living earth. The laws of physics and chemistry respond to the change in the carbon balance, producing effects that challenge the stability of ecosystems. Living things do not adapt quickly to rapid changes in habitat or living conditions. Those that can move migrate in search of better conditions for survival, and this is already happening in human populations. Disguised as "unrest," millions of people are thrust into a forced exodus as lands are inundated by rising seas, parched to dust by drought, or rent asunder by violent storms. Anger, frustration, rebellion, fear, and despair form the undercurrent of divisive politics and destructive policies. When young people are allowed to speak, they call for a stop to the war on nature. They ask for the right to a future on a living planet, with clean air, fresh water, fertile ground, and the biodiversity of species to support them.

Recognizing the atrocities of humanity committed upon the living earth is the beginning point for coalescing a global response. Out of the moral indignation toward the deliberate destruction of the earth can come the resolve to change direction. The accumulated knowledge and technology now embedded in human civilization has brought on much of the destruction, but also holds the tools for change. Harnessing the passion of those outraged and oppressed to the task of preserving the life support system of the earth is the critical element. The idealist's call to "save the world" is also a call to save ourselves.

Faced with the reality of climate change and pervasive global pollution, such aspirations are no longer trivial caricatures of the liberal fringe but the rallying cry of humanity struggling for survival.

The pathways to a society structured in harmony with natural systems already exist and are being adopted from the grassroots up by individuals and communities such as Pittsburgh. In energy, the convergence of information technology with technologies to harness the already distributed renewable energy resources of the earth offers opportunities for rapid transformation of the economy. Likewise, turning the food production system from a chemical-laden, depleting enterprise to one that regenerates fertile soil also captures carbon and sequesters it in the ground. Turning the production of necessary goods and materials from a linear, synthetic-based system to one designed to recover and regenerate materials without hazard and toxic waste offers vast job opportunities. Uniting the moral commitment to preserve the earth and to establish equity in emerging economies will unleash the creative potential of human ingenuity. Comprehensive global aspirations must come down to the practical matters of everyday life in communities. Empowering people to adopt changes and lead the movement to sustainability is the surest way to success.

Major transitions can happen more rapidly when the laws and regulations of society smooth the way. There is a two-hundred-year legacy of laws on property rights, business practices, and public policy. Much of this body of existing law protects the current systems of resource extraction and management, rights of corporations, and the allocation of costs through tax policy and subsidies. Changing the laws governing energy, agriculture, and materials has proven to be extremely difficult at the national level. Even when laws are adopted giving new direction, such as the deregulation of telecommunications or energy, implementation and enforcement fall to the states, each of which reacts uniquely. A unifying national spirit has yet to emerge around climate change or pollution. Indeed, these issues remain highly polarized along partisan lines. However, at the city level, and in communities where people are in closer proximity to the issues and the solutions, positive initiatives emerge.

Leadership matters. Pittsburgh has the benefit of a leader who has raised the hopes of the people with his aspiration. In his inaugural address, Mayor Peduto said:

> It is now our moment, our opportunity—indeed, it is our duty—to create the next Pittsburgh: A city that gleams not only with the lights and glass of our towers, but one that also glows with the hope of revived neighborhoods that thrive in a renewed sense of possibility for every child born here. A city that is more than a place to be from, but rather, a place where lives can be lived to their fullest measure. A city where our children graduate from school and stay, perfect our next economy, raise their families, and hand the next generation a greater city still.[1]

But beyond the direct actions of government, a great upsurge in collaborative spirit flows through Pittsburgh that allows creative interactions to move forward. Public/private partnerships and new ways of bringing resources to problems are being tested. The goal of sustainability and resilience in the face of challenge is clearer when articulated by a strong leader. But it only has credibility when it resonates within the hearts of the people also.

In this city, at the nexus of the fossil era and the sustainable future, the process of realigning land, labor, and capital proceeds steadily. Land use for buildings, roads, and infrastructure still faces the loss of natural cover. But the increasing attention to the benefits of green infrastructure for stormwater management has improved planning, especially in publicly owned spaces. And support for urban agriculture increases each year enabled by supportive zoning changes. Revamping the transportation system provides another platform for dialogue over social justice issues of access, equity, and community integrity. Making the city people-centered rather than vehicle-centered means removing some of the concrete barriers that isolated communities from each other.

Labor presents a more acute challenge. The emergent employment of the new economy will require changes in the workforce, changes that can be slowed or accelerated by the laws and regulations of the economy. Traditional labor unions have been slow to support sustainability and action on climate change. Every job is seen as a good job, especially in traditional endeavors like coal, oil, and gas development and infrastructure. New job opportunities in renewable energy, green chemistry, and regenerative agriculture are not dominant in the current labor market, but hold great potential. The future in a renewable-powered and sustainable economy is not drawn with enough certainty to displace fears over the loss of jobs in the fossil industries. In Pittsburgh, the boom phase of the Marcellus Shale development distracts the labor movement from more forward-looking options. The labor/environment nexus is a rough edge in the transition. The ethical concerns of fairness to workers and respect for their needs is dampened by the pressure for economic growth, even when that growth drives the carbon balance even further into damaging options for the future. The ethical debate has barely begun over balancing the needs of tomorrow with the just treatment of present workers.

Ultimately, the battle over ethics comes down to decisions about where to make investments. The student-initiated Divest from Fossil Fuels movement is growing in Pittsburgh and elsewhere. Discussions in corporate boardrooms, city councils, and among college trustees and individual financial advisors raise the issue of the fate of the future. The implications for shifting investment from fossil fuels to other more sustainable options are still in early stages of definition. But divestment tied to the moral issue of obligations to the future can produce momentum for change. The debate about the need for preserving options for the future runs through strategic planning sessions at many levels. Discussions about investment proceed over such matters as pension fund stability, municipal infrastructure investments, and business development decisions.

The future of young people is being shaped by decisions of their elders. But many young leaders are engaged in this struggle to shape a sustainable future. Pittsburgh is embarked on a transformation once more, leading a path forward based on the resilience, intelligence, and dedication of its people, institutions, and leaders, present and future. Voices of young leaders bring hope and passion to shaping a new age.

Consensus develops slowly, and rarely along clear lines of agreement. A wider commitment to ethical actions that protect the most vulnerable people can make the necessary changes easier to accept. Recognizing that each individual has played a part in creating the problem also has the corollary that each person can have a part in the solution. Empowering people to take responsibility for their role in using the earth's resources opens many possibilities. Shifting consumption patterns and eating patterns can make an enormous difference in the burden on natural systems. A twenty-first-century ethic of living in harmony with nature as a valued virtue can accelerate choices for resource-conserving habits like eating less meat and using less "stuff." Each person makes a difference; each person matters; each voice counts. The critical need to preserve the living Earth, our only home, will require millions of people to speak out for the rights of all living things, for the unborn of future generations, to give voice to the Earth. The collective power of voices united in one purpose can move Congress. The pathways to a sustainable future lie open.

NOTES

INTRODUCTION

Epigraph: Rachel L. Carson, testimony, hearings before the US Senate Subcommittee on Reorganization and International Organizations of the Committee on Government Operations, "Interagency Coordination in Environmental Hazards (Pesticides)," June 4, 1963, 88th Congress, 1st sess. (Washington, DC: Government Printing Office, 1964).

1. President Jimmy Carter awarded a posthumous Presidential Medal of Freedom to Rachel Carson on June 9, 1980. Carter, "Presidential Medal of Freedom Remarks at the Presentation Ceremony," June 9, 1980, American Presidency Project, accessed Jan. 30, 2017, http://www.presidency.ucsb.edu/ws/?pid=45389.
2. Joel Tarr, ed., *Devastation and Renewal: An Environmental History of Pittsburgh and Its Region* (Pittsburgh, PA: University of Pittsburgh Press, 2003), 3–4.
3. Intergovernmental Panel on Climate Change, *Climate Change 2007: Synthesis Report*, Contribution of Working Groups I, II, and III to the Fourth Assessment Report of the Intergovernmental Panel on Climate Change, IPCC

Plenary 27, Valencia, Spain, Nov. 12–17, 2007, 36, http://www.ipcc.ch/pdf/assessment-report/ar4/syr/ar4_syr.pdf; Åke Bergman, Jerrold J. Heindel, Susan Jobling, Karen A. Kidd, and R. Thomas Zoeller, *State of the Science of Endocrine Disrupting Chemicals, 2012* (Geneva, Switzerland: United Nations Environment Programme and the World Health Organization, 2013), vii–ix.

4. J. D. Hays, John Imbrie, and N. J. Shackleton, "Variations in the Earth's Orbit: Pacemaker of the Ice Ages," *Science* 194, no. 4270 (1976): 1121–32, doi:10.1126/science.194.4270.1121.
5. Edward A. Keller, *Environmental Geology*, 9th ed. (Upper Saddle River, NJ: Pearson Prentice Hall Education, 2011), 485–86.
6. Judith L. Lean and David H. Rind, "How Natural and Anthropogenic Influences Alter Global and Regional Surface Temperatures: 1889 to 2006," *Geophysical Research Letters* 35, no. 18 (2008): L18701, doi:10.1029/2008GL034864.
7. Franklin Hadley Cocks, "Global Warming vs the Next Ice Age," *MIT Technology Review*, Dec. 21, 2009, accessed Mar. 24, 2017, https://www.technologyreview.com/s/416786/global-warming-vs-the-next-ice-age/.
8. US Environmental Protection Agency, "Carbon Pollution Emission Guidelines for Existing Stationary Sources: Electric Utility Generating Units," final rule (Washington, DC: US Government Publishing Office, 2015), *Federal Register* 80, no. 205 (Oct. 23, 2015): 64663–65, https://www.gpo.gov/fdsys/pkg/FR-2015-10-23/pdf/2015-22842.pdf.
9. Thomas F. Stocker, Dahe Qin, Gian-Kasper Plattner, Melinda M. B. Tignor, Simon K. Allen, Judith Boschung, Alexander Nauels, Yu Xia, Vincent Bex, and Pauline M. Midgley, eds., *Climate Change 2013: The Physical Science Basis: Working Group I Contribution to the Fifth Assessment Report of the Intergovernmental Panel on Climate Change: Summary for Policymakers* (Geneva, Switzerland: Intergovernmental Panel on Climate Change, 2013), 5, https://www.ipcc.ch/pdf/assessment-report/ar5/wg1/WGIAR5_SPM_brochure_en.pdf.
10. John Eddy, *A New Sun: The Solar Results from Skylab*, ed. Rein Ise (Washington, DC: National Aeronautics and Space Administration, 1979), 37, http://history.nasa.gov/SP-402/sp402.htm.
11. Stocker et al., *Climate Change 2013*, 6.
12. World Economic Forum, *Global Risks 2014*, 9th ed. (Geneva, Switzerland: World Economic Forum, 2014), 14. This report and an interactive data platform are available at www.weforum.org/risks.

13. Intergovernmental Panel on Climate Change, *Climate Change 2014: Synthesis Report.* Contribution of Working Groups I, II and III to the Fifth Assessment Report of the Intergovernmental Panel on Climate Change, ed. Rajendra K. Pachauri and Leo A. Meyer (Geneva, Switzerland: Intergovernmental Panel on Climate Change, 2014), https://www.ipcc.ch/report/ar5/syr/.
14. National Research Council, *Climate Stabilization Targets: Emissions, Concentrations, and Impacts over Decades to Millennia* (Washington, DC: National Academies Press, 2011), 24–25.
15. US Department of Defense, *2014 Climate Change Adaptation Roadmap* (Alexandria, VA: Office of the Deputy Under Secretary of Defense for Installations and Environment, 2014), ii, http://www.acq.osd.mil/eie/Downloads/CCARprint_wForward_e.pdf.
16. US Department of Defense, *2014 Climate Change Adaptation Roadmap*, 1–3.
17. Gina McCarthy, Environmental Protection Agency administrator, quoted in "EPA Proposes First Guidelines to Cut Carbon Pollution from Existing Power Plants," news release, Environmental Protection Agency, Washington, DC, June 2, 2014, http://yosemite.epa.gov/opa/admpress.nsf/bd4379a92ceceeac8525735900400c27/5bb6d20668b9a18485257ceb00490c98!OpenDocument.
18. Conrad Schneider and Jonathan Banks, *The Toll from Coal: An Updated Assessment of Death and Disease from America's Dirtiest Energy Source* (Boston: Clean Air Task Force, 2010), 10–12.
19. National Research Council, *Hidden Costs of Energy: Unpriced Consequences of Energy Production and Use* (Washington, DC: National Academies Press, 2010), 104–42.
20. Theo Colborn, Dianne Dumanoski, and John Peterson Myers, *Our Stolen Future: Are We Threatening Our Fertility, Intelligence, and Survival?: A Scientific Detective Story* (New York: Penguin, 1997), xvi.
21. US Environmental Protection Agency, *2014 TRI National Analysis: Executive Summary* (Washington, DC: Environmental Protection Agency, 2016), ii, https://www.epa.gov/sites/production/files/2016-01/documents/2014-tri-na-exec-summary.pdf.
22. Charles McCollester, *The Point of Pittsburgh: Production and Struggle at the Forks of the Ohio* (Pittsburgh, PA: Battle of Homestead Foundation, 2008), 85–89.
23. Rachel Carson, *The Sense of Wonder* (New York: HarperCollins, 1998), 100.

CHAPTER 1. LISTENING TO THE VOICE OF THE EARTH

Epigraphs: Rachel Carson, "On the Pollution of the Environment," in *Lost Woods: The Discovered Writing of Rachel Carson*, ed. Linda J. Lear (Boston: Beacon Press, 1998), 228; World Commission on the Ethics of Scientific Knowledge and Technology, *The Ethical Implications of Global Climate Change: Report by the World Commission on the Ethics of Scientific Knowledge and Technology* (Paris: United Nations Educational, Scientific and Cultural Organization, 2010), 13; Paul R. Ehrlich, *Human Natures: Genes, Cultures, and the Human Prospect* (Washington, DC: Island Press, 2000), 279; Barbara Ward and René J. Dubos, *Only One Earth: The Care and Maintenance of a Small Planet* (New York: W. W. Norton, 1972), 114; Carson, "On the Pollution of the Environment," 242–43; CBS Reports, "The Silent Spring of Rachel Carson," aired Apr. 13, 1963, http://www.downvids.net/cbs-reports-the-silent-spring-of-rachel-carson-548117.html.

1. Angela Gugliotta, "How, When, and for Whom Was Smoke a Problem in Pittsburgh?," in *Devastation and Renewal: An Environmental History of Pittsburgh and Its Region*, ed. Joel A. Tarr (Pittsburgh, PA: University of Pittsburgh Press, 2003), 114.
2. Christian Nellemann and Emily Corcoran, eds., *Dead Planet, Living Planet: Biodiversity and Ecosystem Restoration for Sustainable Development: A Rapid Response Assessment* (Nairobi, Kenya: United Nations Environment Programme, 2010), 62.
3. Roxanne Dunbar-Ortiz, *An Indigenous Peoples' History of the United States* (Boston: Beacon Press, 2014), 17.
4. Kevin Hillstrom, *U.S. Environmental Policy: A Documentary History* (Washington, DC: CQ Press, 2010), 2–6.
5. Hillstrom, *U.S. Environmental Policy*, 63.
6. US Geological Survey, "Coal-Mine-Drainage Projects in Pennsylvania," *Pennsylvania Water Science Center*, last modified Oct. 26, 2010, http://pa.water.usgs.gov/projects/energy/amd/.
7. S. R. Jennings, D. R. Neuman, and P. S. Blicker, *Acid Mine Drainage and Effects on Fish Health and Ecology: A Review* (Bozeman, MT: Reclamation Research Group for US Fish and Wildlife Service, 2008); see also Trout Unlimit-

ed, "Trout Unlimited Celebrates Dramatic Recovery in the West Branch of the Susquehanna Watershed," press release, Oct. 24, 2011, http://www.tu.org/press_releases/2011/trout-unlimited-celebrates-dramatic-revocery-in-the-west-branch-susquehanna-wate.

8. US Environmental Protection Agency Office of Research and Development, National Center for Environmental Assessment, *The Effects of Mountaintop Mines and Valley Fills on Aquatic Ecosystems of the Central Appalachian Coalfields (2011 Final)*, EPA/600/R-09/138F (Washington, DC: US Environmental Protection Agency, 2011), 19.
9. Krystof Obidzinski, Rubeta Andriani, Heru Komarudin, and Agus Andrianto, "Environmental and Social Impacts of Oil Palm Plantations and Their Implications for Biofuel Production in Indonesia," *Ecology and Society* 17, no. 1 (2010): 25.
10. Chris D. Thomas, Alison Cameron, Rhys E. Green, Michel Bakkenes, Linda J. Beaumont, Yvonne C. Collingham, Barend F. N. Erasmus, Marinez Ferreira de Siqueira, Alan Grainger, Lee Hannah, Lesley Hughes, Brian Huntley, Albert S. van Jaarsveld, Guy F. Midgley, Lera Miles, Miguel A. Ortega-Huerta, A. Townsend Peterson, Oliver L. Phillips, and Stephen E. Williams, "Extinction Risk from Climate Change," *Nature* 427, no. 6970 (Jan. 8, 2004): 145–48.
11. Jeffrey K. McKee, Paul W. Sciulli, C. David Fooce, and Thomas A. Waite, "Forecasting Biodiversity Threats Due to Human Population Growth," *Biological Conservation* 115, no. 1 (2004): 161–64.
12. Jeffrey D. Sachs, *Common Wealth: Economics for a Crowded Planet* (New York: Penguin, 2008), 18.
13. Stephen M, Gardiner, "A Perfect Moral Storm: Climate Change, Intergenerational Ethics, and the Problem of Moral Corruption," *Environmental Values* 15 (2006): 397–413.
14. Hervé Le Treut, Richard Somerville, Ulrich Cubasch, Yihui Ding, Cecilie Mauritzen, Abdalah Mokssit, Thomas Peterson, Michael Prather, et al., "Historical Overview of Climate Change Science," in *Climate Change 2007: The Physical Science Basis*, Contribution of Working Group I to the Fourth Assessment Report of the Intergovernmental Panel on Climate Change, ed. Susan Solomon, Dahe Qin, Martin Manning, Melinda Marquis, Kristen Averyt, Melinda M. B. Tignor, Henry LeRoy Miller, and Zhenlin Chen (New York: Cambridge University Press, 2007), 93–128.

15. Bill McKibben, "Falling Short on Climate in Paris," *New York Times*, Dec. 13, 2015, http://www.nytimes.com/2015/12/14/opinion/falling-short-on-climate-in-paris.html?_r=0.
16. Marvin S. Soroos, "Preserving the Atmosphere as a Global Commons," *Environment* 40, no. 2 (1998): 32.
17. Intergovernmental Panel on Climate Change, *Climate Change 2007: Synthesis Report—Summary for Policymakers*, Summary of the IPCC Plenary 27, Valencia, Spain, Nov. 12–17, 2007 (Geneva, Switzerland: United Nations, 2007), 2–4, https://www.ipcc.ch/pdf/assessment-report/ar4/syr/ar4_syr_spm.pdf.
18. Soroos, "Preserving the Atmosphere," 6–13.
19. Timothy Gore, *Extreme Carbon Inequality*, Oxfam International, Media Briefing, Dec. 2, 2015, 4, 12, https://www.oxfam.org/sites/www.oxfam.org/files/file_attachments/mb-extreme-carbon-inequality-021215-en.pdf.
20. E. Lynn Usery, *Modeling Sea-Level Rise Effects on Population Using Global Elevation and Land-Cover Data*, US Geological Survey, 2007, http://cegis.usgs.gov/pdf/aag-2007.pdf.
21. United Nations Framework Convention on Climate Change, *Report of the Conference of the Parties on Its Twenty-first Session, Held in Paris from 30 November to 15 December, 2015* (Paris: United Nations, 2016), 35–36, http://unfccc.int/resource/docs/2015/cop21/eng/10.pdf.
22. United Nations Framework Convention on Climate Change, Adoption of the Paris Agreement, Draft Decision/CP21 (Paris: United Nations, 2015), 2, https://unfccc.int/resource/docs/2015/cop21/eng/l09r01.pdf.
23. Richard C. Somerville, "The Ethics of Climate Change," *Yale Environment 360*, June 2, 2008, accessed Jan. 31, 2017, http://e360.yale.edu/feature/the_ethics_of_climate_change/1365.
24. Gardiner, "Perfect Moral Storm," 401.
25. Linda R. Berg, Mary Catherine Hager, and David M. Hassenzahl, *Visualizing Environmental Science*, 3rd ed. (Hoboken, NJ: John Wiley, 2011), 106–7.
26. World Meteorological Association, "The State of Greenhouse Gases in the Atmosphere Based on Global Observations through 2013," WMO Greenhouse Gas Bulletin no. 10, Nov. 2014.
27. Philip Cafaro, "Thoreau, Leopold, and Carson: Toward an Environmental Virtue Ethics," in *Environmental Virtue Ethics*, ed. Ronald Sandler and Philip Cafaro (Lanham, MD: Rowman and Littlefield, 2005), 31–33.

28. Patrick G. Derr and Edward M. McNamara, *Case Studies in Environmental Ethics* (Lanham, MD: Rowman & Littlefield, 2003), xv–xxi.
29. Cafaro, "Thoreau, Leopold, and Carson," 39–40.
30. Cafaro, "Thoreau, Leopold, and Carson," 32.
31. Ehrlich, *Human Natures*, 319.
32. Paul R. Ehrlich and John P. Holdren, "Impact of Population Growth," *Science* 171, no. 3977 (1971): 1212–17, doi:10.1126/science.171.3977.1212.
33. United Nations Department of Economic and Social Affairs, *World Population Prospects: The 2002 Revision*, vol. 3, *Analytical Report* (New York: United Nations, 2004), 2, http://www.un.org/esa/population/publications/wpp2002/WPP2002_VOL_3.pdf.
34. United Nations Department of Economic and Social Affairs, *World Population Prospects: The 2002 Revision*, 3:3.
35. "Prepared Philippine Statement on Paris Agreement at the 21st Conference of Parties (COP21)," delivered by Secretary Emmanuel M. de Guzman, Dec. 12, 2015, http://newsinfo.inquirer.net/747138/full-text-paris-agreement-philippine-statement-cop21#ixzz4AirwIWdP.
36. Francis, *Laudato Si': On Care for Our Common Home* (Vatican City, Italy: Libreria Editrice Vaticana, 2015), par. 63, accessed Jan. 30, 2017, http://w2.vatican.va/content/francesco/en/encyclicals/documents/papa-francesco_20150524_enciclica-laudato-si.html.
37. Islamic Foundation for Ecology and Environmental Sciences, *Islamic Declaration on Global Climate Change*, drafted at International Islamic Climate Change Symposium, Istanbul, Turkey, Aug. 18, 2015, http://www.ifees.org.uk/wp-content/uploads/2016/10/climate_declarationmMWB.pdf.
38. Thich Nhat Hanh, "The Bells of Mindfulness," in *Spiritual Ecology: The Cry of the Earth*, ed. Llewellyn Vaughan-Lee (Point Reyes, CA: Golden Sufi Center, 2013), 25–28.
39. "The Proposals of 'Peoples Agreement' in the Texts for United Nations Negotiations," *World People's Conference on Climate Change and the Rights of Mother Earth*, Aug. 16, 2010, accessed July 16, 2016, https://pwccc.wordpress.com/2010/08/16/the-proposals-of-%E2%80%9Cpeoples-agreement%E2%80%9D-in-the-texts-for/.
40. "Proposals of 'Peoples Agreement.'"
41. Karl-Henrik Robèrt, "The Natural Step as a Framework for Strategic Sustainable Development," Rachel Carson Legacy Conference, "Challenging Mar-

cellus Shale: Consequences and Alternatives," Carnegie Mellon University, Mellon Institute, Sept. 24, 2010.

42. Patricia M. DeMarco, "Rachel Carson's Environmental Ethic—A Guide for Global Systems Decision Making," *Journal of Cleaner Production* 140 (2017): 58, doi:10.1016/j.jclepro.2015.03.058.
43. Albert Schweitzer correspondence with Rachel Carson, Mar. 16, 1963, Rachel Carson Council Collection, National Conservation Training Center, United States Fish and Wildlife Service Museum and Archives.
44. Linda J. Lear, *Rachel Carson: Witness for Nature*, 2nd ed. (Boston: Mariner Books, 2009), 438.
45. Rachel L. Carson, *The Sea around Us*, illus. comm. ed. (New York: Oxford University Press, 2003), xi, xiii.
46. Carson, "On the Pollution of the Environment," 228–45.
47. Lear, *Rachel Carson*, 16.
48. Mark E. Dixon, producer, and Patricia DeMarco, executive producer, *The Power of One Voice: A 50-Year Perspective on the Life of Rachel Carson* (Pittsburgh, PA: Blue Lens, LLC, Steeltown Entertainment, 2015), www.PowerofOneVoiceFilm.com.
49. William Peduto, "The Next Pittsburgh," inaugural address, *Pittsburgh Post-Gazette*, Jan. 6, 2014, http://www.post-gazette.com/local/city/2014/01/06/Trasncript-Bill-Peduto-s-inaugural-speech/stories/201401060154.
50. Diana Nelson Jones, "Pittsburgh Recognized as Biophilic City," *Pittsburgh Post-Gazette*, Sept. 17, 2016, accessed Oct. 10, 2016, http://www.post-gazette.com/local/city/2016/09/17/Pittsburgh-s-cap-gets-another-feather-with-induction-into-Biophilic-Cities-network-an-effort-to-integrate-nature-into-daily-city-life/stories/201609170031.
51. DeMarco, "Rachel Carson's Environmental Ethic," 58.
52. World Commission on the Ethics of Scientific Knowledge and Technology, "History of the Precautionary Principle," in *The Precautionary Principle* (Paris: United Nations Educational, Scientific and Cultural Organization, 2005), 9, http://unesdoc.unesco.org/images/0013/001395/139578e.pdf.
53. Commission of the European Communities, "Communication from the Commission on the Precautionary Principle" (Brussels: Commission of the European Communities, 2000), 10, http://eur-lex.europa.eu/LexUriServ/LexUriServ.do?uri=COM:2000:0001:FIN:EN:PDF.
54. Karl-Henrick Robèrt, "Tools and Concepts for Sustainable Development,

How Do They Relate to a Framework for Sustainable Development, and to Each Other?," *Journal of Cleaner Production* 8, no. 3 (2000): 243–54.

55. "Case Studies," *The Natural Step*, accessed July 16, 2016, http://www.thenaturalstep.org/cases/.
56. Karl-Henrick Robèrt, "Integrating Sustainability into Business Strategy and Operations: Applying The Natural Step Approach and Framework and Back-Casting from Principles of Sustainability," in *Ants, Galileo, and Gandhi: Designing the Future of Business through Nature, Genius, and Compassion*, ed. Sissel Waage (Sheffield, UK: Greenleaf, 2003), 61–80.
57. Karl-Henrick Robèrt, "Integrating Sustainability," 61–80; see also Karl-Henrik Robèrt, Sven Boren, Henrik Ny, and Goran Borman, "A Strategic Approach to Sustainable Transport System Development. Part 1: Attempting A Generic Community Planning Process Model," *Journal of Cleaner Production*, 140 (2017): 56.

CHAPTER 2. PRESERVING THE LIVING EARTH

Epigraphs: Edward O. Wilson, *The Diversity of Life* (Cambridge, MA: Harvard University Press, 1992), 347; Rachel Carson, "On the Pollution of the Environment," in *Lost Woods: The Discovered Writing of Rachel Carson*, ed. Linda J. Lear (Boston: Beacon Press, 1998), 227; Rachel Carson, submission to *Outdoor Life* competition, Oct. 15, 1946, Lear/Carson Collection, Connecticut College, New London; Rachel Carson, *Guarding Our Wildlife Resources*, Conservation in Action Series, no. 5 (Washington, DC: US Fish and Wildlife Service, 1943), 1.

1. To learn more about the National Wildlife Refuge management programs as well as national wildlife refuge management practices and other locations, visit US Fish and Wildlife Service, *National Wildlife Refuge System*, http://www.fws.gov/refuges.
2. Rachel L. Carson, *Chincoteague: A National Wildlife Refuge*, Conservation in Action Series, no. 1 (Washington, DC: US Department of the Interior, Fish and Wildlife Service, 1947), www.fws.gov/rachelcarson.
3. Lou Hinds, director, Chincoteague National Wildlife Refuge, author interview, Aug. 11, 2011.
4. For information about the Chincoteague National Wildlife Refuge, see US

Fish and Wildlife Service, *Chincoteague National Wildlife Refuge*, https://www.fws.gov/refuge/chincoteague/.

5. *Chincoteague Chamber of Commerce*, www.chincoteaguechamber.com.
6. Personal conversations with several people in the town who wished to remain anonymous, Aug. 11–15, 2011.
7. Marguerite Henry, *Misty of Chincoteague* (Chicago: Rand McNally, 1947); film: *Misty*, dir. James B. Clark, perf. David Ladd, Arthur O'Connell (1961, Port Washington, NY: Koch Vision, 2008), DVD.
8. Hinds, author interview, Aug. 11, 2011.
9. Paul Falkowski, "Ocean Science: The Power of Phytoplankton," *Nature* 483, no. 7387 (Mar. 1, 2012): S17–S20.
10. Christian Nellemann and Emily Corcoran, eds., *Dead Planet, Living Planet: Biodiversity and Ecosystem Restoration for Sustainable Development: A Rapid Response Assessment* (Nairobi, Kenya: United Nations Environment Programme, 2010), 65.
11. John Eddy, *A New Sun: The Solar Results from Skylab*, ed. Rein Ise (Washington, DC: National Aeronautics and Space Administration, 1979), 36, http://history.nasa.gov/SP-402/sp402.htm.
12. For an encyclopedia of life on Earth, see the site *Encyclopedia of Life*, http://www.eol.org. This is the most complete documentation of species available.
13. Wilson, *Diversity of Life*, xi.
14. Eugene P. Odum, *Fundamentals of Ecology*, 2nd ed. (Philadelphia, PA: W. B. Saunders, 1966), 10.
15. Robert Costanza, Ralph d'Arge, Rudolph de Groot, Stephen Farber, Monica Grasso, Bruce Hannon, Karin Limburg, Shahid Naeem, Robert V. O'Neill, Jose Paruelo, Robert G. Raskin, Paul Sutton, and Marjan van den Belt, "The Value of the World's Ecosystem Services and Natural Capital," *Nature* 387, no. 6630 (1997): 253–54.
16. Millennium Ecosystem Assessment Board, *Ecosystems and Human Well-Being: Synthesis* (Washington, DC: Island Press, 2005), 10–12, http://www.millenniumassessment.org/documents/document.356.aspx.pdf.
17. Nellemann and Corcoran, *Dead Planet*, 12.
18. Secretariat of the Convention on Biological Diversity, *Global Biodiversity Outlook 3* (Montreal: Secretariat of the Convention on Biological Diversity, 2010), 15, https://www.cbd.int/doc/publications/gbo/gbo3-final-en.pdf.

19. Secretariat of the Convention on Biological Diversity, *Global Biodiversity Outlook* 3, 32.
20. Secretariat of the Convention on Biological Diversity, *Global Biodiversity Outlook* 3, 58.
21. Bradley J. Cardinale, J. Emmett Duffy, Andrew Gonzalez, David U. Hooper, Charles Perrings, Patrick Venail, Anita Narwani, Georgina M. Mace, David Tilman, David A. Wardle, Ann P. Kinzig, Gretchen C. Daily, Michel Loreau, James B. Grace, Anne Larigauderie, Diane S. Srivastava, and Shahid Naeem, "Biodiversity Loss and Its Impact on Humanity," *Nature* 486, no. 7401 (2012): 59–67, doi:10.1038/nature1114807.
22. Charles McCollester, *The Point of Pittsburgh: Production and Struggle at the Forks of the Ohio* (Pittsburgh, PA: Battle of Homestead Foundation, 2008), 193–95.
23. McCollester, *Point of Pittsburgh*, 199–202.
24. McCollester, *Point of Pittsburgh*, 218–25.
25. Rob Nixon, *Slow Violence and the Environmentalism of the Poor* (Cambridge, MA: Harvard University Press, 2011), 1.
26. Nixon, *Slow Violence*, 14.
27. On average, Americans recycled and composted 1.51 pounds out of the individual waste generation rate of 4.38 pounds per person per day. US Environmental Protection Agency, *Municipal Solid Waste Generation, Recycling, and Disposal in the United States: Facts and Figures for 2012*, accessed July 18, 2016, https://www.epa.gov/sites/production/files/2015-09/documents/2012_msw_fs.pdf.
28. Florian Thevenon, Chris Carroll, and João Sousa, eds., *Plastic Debris in the Ocean: The Characterization of Marine Plastics and Their Environmental Impacts, Situation Analysis Report* (Gland, Switzerland: International Union for Conservation of Nature and Natural Resources, 2014), 11–13, https://portals.iucn.org/library/sites/library/files/documents/2014-067.pdf.
29. Thevenon, Carroll, and Sousa, *Plastic Debris*, 27.
30. Claire Groden, "Report: Plastic Pollution in the Ocean Is Reaching Crisis Levels," *Fortune*, Oct. 1, 2015, http://fortune.com/2015/10/01/ocean-plastic-pollution/.
31. Exercise given to University of Pittsburgh undergraduate students in the Environmental Studies Program course GEOL 1904, spring 2009, taught by Patricia M. DeMarco.

32. See details about the film and the YERT Project here: *YERT: The Film*, http://www.yert.com/film.php.
33. Arthur Grube, David Donaldson, Timothy Kiely, and La Wu, "Pesticide Industry Sales and Usage: 2006 and 2007 Market Estimates," US EPA Report EPA 733-R-11-001 (Washington, DC: US Environmental Protection Agency, 2011), https://www.epa.gov/sites/production/files/2015-10/documents/market_estimates2007.pdf. Ninety-eight percent of herbicides and ninety-four percent of pesticides applied to domestic gardens kill nontarget species. Annually, 1.1 billion pounds of herbicides and pesticides are applied.
34. For native planting guides across the country see the US Fish and Wildlife Service website, www.fws.gov; the Garden Club of America website, www.gcamerica.org; and the Agricultural Extension Service of USDA, National Wildlife Federation Backyard Wildlife Habitat program.
35. The City of Pittsburgh commitment in the Rachel Carson Legacy Challenge, 2007, accessed Mar. 9, 2017, https://rachelcarsonhomestead.org/rachel-carson-legacy/legacy-challenge-commitments/.
36. The Allegheny Conference on Regional Development, headquartered in Pittsburgh, emphasizes improved waterways, trails, and environmental quality as important priorities. See Pittsburgh Regional Alliance, "Quality of Place," accessed Feb. 18, 2016, http://www.pittsburghregion.org/why/quality-of-place/.
37. Pittsburgh Botanic Garden, "History of the Land," accessed Mar. 16, 2017, http://pittsburghbotanicgarden.org/about-us-2/history-of-the-land/.
38. Pittsburgh Botanic Garden, "History of the Land."
39. Pittsburgh Botanic Garden, "History of the Garden," accessed Mar. 16, 2017, http://pittsburghbotanicgarden.org/about-us-2/history/.
40. Pittsburgh Botanic Garden, "History of the Garden."
41. Pittsburgh Botanic Garden, "Land Reclamation on the Garden Site," accessed Apr. 12, 2017, http://pittsburghbotanicgarden.org/reclamation/.
42. Justin Coyne, "Pittsburgh Botanic Garden Celebrates Grand Opening after Two Decades of Work," *Pittsburgh Business Times*, Aug. 1, 2014, http://www.bizjournals.com/pittsburgh/news/2014/08/01/pittsburgh-botanic-garden-celebrates-grand-opening.html.
43. Madasyn Czebiniak, "Pittsburgh Botanic Garden Brims with Life after Abandoned Mine Cleanup," *Trib Live*, Oct. 24, 2016, http://triblive.com/news/allegheny/11347985-74/garden-mine-pond.

44. Czebiniak, "Pittsburgh Botanic Garden."
45. US Geological Survey, "Coal-Mine-Drainage Projects in Pennsylvania," *Pennsylvania Water Science Center*, page last modified Oct. 26, 2010, accessed Jan. 12, 2016, http://pa.water.usgs.gov/projects/energy/amd/.
46. Andrew S. McElwaine, "Slag in the Park," in *Devastation and Renewal: An Environmental History of Pittsburgh and Its Region*, ed. Joel A. Tarr (Pittsburgh, PA: University of Pittsburgh Press, 2003), 174, 189.
47. McElwaine, "Slag in the Park," 190–91.
48. The Studio for Creative Inquiry is a program in the School of Fine Arts at Carnegie Mellon University. "Nine Mile Run Greenway Project," *Frank-Ratchye Studio for Creative Inquiry*, accessed Jan. 31, 2017, http://studioforcreativeinquiry.org/projects/nine-mile-run-greenway-project.
49. Nine Mile Run Watershed Association, "History," accessed Jan. 31, 2017, www.ninemilerun.org/history.
50. Nine Mile Run Watershed Association, *Ample Opportunity: A Community Dialogue*, Final Report: Nine Mile Run Greenway Project, Feb. 1998, http://nmr.collinsandgoto.com/publications/content/ampleopp/content.htm.
51. Nine Mile Run Watershed Association, *2013 State of the Watershed* (Pittsburgh, PA: Nine Mile Run Watershed Association), fig. 3, p. 4. See also the association website at www.ninemilerun.org.
52. Allegheny County Sanitation Authority operates sewer treatment facilities along the Monongahela River.
53. Nine Mile Run Watershed Association, *State of the Watershed Report 2009*, 1.
54. Nine Mile Run Watershed Association, *State of the Watershed Report 2009*, 3.
55. Nine Mile Run Watershed Association, "Monitoring Committee and Collaborative Partners," *State of the Watershed 2010* (Pittsburgh, PA: Nine Mile Run Watershed Association, 2011), 15.
56. Nine Mile Run Watershed Association, "Mission and Community," *State of the Watershed 2009*, 17.
57. Personal communication with Brenda Smith, executive director, Nine Mile Run Watershed Association, Jan. 22, 2014.
58. Borough of Forest Hills, "General Information about Forest Hills Borough," accessed Apr. 8, 2016, http://www.foresthillspa.org/fhpa-about.html.
59. John Mburu, Lars Gerard Hein, Barbara Gemmill, and Linda Collette, *Tools for Conservation and Use of Pollination Services—Valuation of Pollination*

Services: A Review of Methods (Rome: Food and Agriculture Organization of the United Nations, 2006), 24.

CHAPTER 3. INSPIRING A SENSE OF WONDER

Epigraphs: Rachel L. Carson, *The Sense of Wonder* (New York: HarperCollins, 1998), 54; Rachel Carson, "The Real World around Us," speech to Theta Sigma Phi, Sorority of Women Journalists, 1954, in *Lost Woods: The Discovered Writing of Rachel Carson*, ed. Linda J. Lear (Boston: Beacon Press, 1998), 160; Janine M. Benyus, *Biomimicry: Innovation Inspired by Nature* (New York: HarperCollins, 1997), 9; Mark L. Winston, *Nature Wars: People vs. Pests* (Cambridge, MA: Harvard University Press, 1997), viii.

1. Dan Ashe, *Statement of Dan Ashe, Director, U.S. Fish and Wildlife Service before the U.S. House of Representatives Appropriations Committee, Interior Subcommittee, Regarding the Fiscal Year 2016 Budget, March 17, 2015*, http://docs.house.gov/meetings/AP/AP06/20150317/103113/HHRG-114-AP06-Wstate-AsheD-20150317.pdf.
2. Richard Louv, *Last Child in the Woods: Saving Our Children from Nature-Deficit Disorder* (Chapel Hill, NC: Algonquin Books of Chapel Hill, 2008), 19.
3. Shobha Srinivasan, Liam R. O'Fallion, and Allen Dearry, "Creating Healthy Communities, Healthy Homes, Healthy People: Initiating a Research Agenda on the Built Environment and Public Health," *American Journal of Public Health* 93, no. 9 (2003): 1446–50.
4. US Department of Health and Human Services, Health Resources and Services Administration, Maternal and Child Health Bureau, *Child Health USA 2014* (Rockville, MD: US Department of Health and Human Services, 2015), 41.
5. Office of the United Nations High Commissioner for Human Rights, *Convention on the Rights of the Child. General Assembly Resolution 44/25*, Nov. 20, 1989, http://www.un.org/documents/ga/res/44/a44r025.htm.
6. Steven R. Kellert, *Public Attitudes toward Critical Wildlife and Natural Habitat Issues* (Washington, DC: Department of the Interior, US Fish and Wildlife Service, 1979), 138.
7. Edward O. Wilson, *Naturalist* (Washington, DC: Island Press, 1994), 362.

8. John M. Zelenski and Elizabeth K. Nisbet, “Happiness and Feeling Connected: The Distinct Role of Nature Relatedness,” *Environment and Behavior* 46, no. 1 (2014): 3–23.
9. Gallup Poll of Public Opinion on the Environment from 2001 to 2015, accessed Apr. 8, 2016, http://www.gallup.com/poll/1615/environment.aspx.
10. Louis J. Guillette, “Environmental Contaminants and Health: Fifty Years of Lessons from Wildlife,” Perspectives on *Silent Spring* at 50 Conference, Chatham University, Pittsburgh, PA, May 12, 2012, http://www.chatham.edu/rachelcarson/events/silentspring50/agenda.cfm.
11. David Sobel, *Beyond Ecophobia: Reclaiming the Heart of Nature Education* (Great Barrington, MA: Orion Society, 1996), 14.
12. Nancy M. Wells, “At Home with Nature: Effects of ‘Greenness’ on Children’s Cognitive Functioning,” *Environment and Behavior* 32, no. 6 (2000): 775–95.
13. Ingunn Fjørtoft, “The Natural Environment as a Playground for Children: The Impact of Outdoor Play Activities in Pre-Primary School Children,” *Early Childhood Education Journal* 29, no. 2 (2001): 111–17.
14. Robin C. Moore and Herbert H. Wong, *Natural Learning: The Life of an Environmental Schoolyard: Creating Environments for Rediscovering Nature’s Way of Teaching* (Berkeley, CA: MIG Communications, 1997), 134.
15. Robin Moore, “Compact Nature: The Role of Playing and Learning Gardens on Children’s Lives,” *Journal of Therapeutic Horticulture* 8 (1996): 72–82.
16. Nancy M. Wells and Gary W. Evans, “Nearby Nature: A Buffer of Life Stress among Rural Children,” *Environment and Behavior* 35, no. 3 (2003): 311–30.
17. William Crain, “How Nature Helps Children Develop,” *Montessori Life*, Summer 2001, 22–24.
18. Steven R. Kellert, *Building for Life: Designing and Understanding the Human-Nature Connection* (Washington, DC: Island Press, 2005), 63–90.
19. Andrea Faber Taylor and Frances E. Kuo, “Is Contact with Nature Important for Healthy Child Development? State of the Evidence,” in *Children and Their Environments: Learning, Using and Designing Spaces*, ed. Christopher Spencer and Mark Blades (Cambridge, UK: Cambridge University Press, 2006), 124–40.
20. Frances E. Kuo, *Parks and Other Green Environments: Essential Components of a Healthy Human Habitat*, National Recreation and Parks Association Research Series 2010 (Ashburn, VA: National Recreation and Parks Association, 2010), 25–30, http://www.nrpa.org/uploadedFiles/nrpa.org/

Publications_and_Research/Research/Papers/MingKuo-Research-Paper .pdf.

21. R. S. Ulrich, "View through a Window May Influence Recovery from Surgery," *Science* 224, no. 4647 (1984): 420–21.
22. J. Maas, R. A. Verheij, S. de Vries, P. Spreeuwenberg, F. G. Schellevis, and P. P. Groenewegen, "Morbidity Is Related to a Green Living Environment," *Journal of Epidemiology and Community Health* 63, no. 12 (2009): 962.
23. Maas, Verheij, de Vries, et al., "Morbidity," 967–70.
24. Jeffrey Koplan, Catharyn Liverman, and Vivica Kraak, eds., *Preventing Childhood Obesity: Health in the Balance* (Washington, DC: National Academies Press, 2005), 81.
25. Edward O. Wilson, "Biophilia and the Conservation Ethic," in *The Biophilia Hypothesis*, ed. Steven R. Kellert and Edward O. Wilson (Washington, DC: Island Press, 1993), 31–42.
26. Zelenski and Nisbet, "Happiness and Feeling Connected," 3–23.
27. Colin A. Capaldi, Raelyne L. Dopko, and John M. Zelenski, "The Relationship between Nature Connectedness and Happiness: A Meta-Analysis," *Frontiers in Psychology*, Sept. 8, 2014, https://doi.org/10.3389/fpsyg.2014.00976.
28. Carson, *Sense of Wonder*, 87.
29. David Cumes, "Nature as Medicine: The Healing Power of the Wilderness," *Alternative Therapies in Health and Medicine* 4, no. 2 (1998): 79–86.
30. The Student Conservation Association of Pittsburgh, accessed Apr. 2, 2016, https://www.thesca.org/serve/program/pittsburgh-pa.
31. John Hayes, "Bug Camp: Rachel Carson Homestead Teaches Kids about Nature," *Pittsburgh Post-Gazette*, July 18, 2010, http://www.post-gazette.com/local/east/2010/07/18/Bug-Camp-Rachel-Carson-Homestead-teaches-kids-about-nature/stories/201007180189.
32. Louv, *Last Child in the Woods*, 36.
33. Rhonda Clements, "An Investigation of the Status of Outdoor Play," *Contemporary Issues in Early Childhood* 5, no. 1 (2004): 68–80.
34. Wilson, *Naturalist*, 56; Rachel Carson, "My Favorite Recreation," in Lear, *Lost Woods*, 12–13.
35. Robin C. Moore, "The Need for Nature: A Childhood Right," *Social Justice* 24, no. 3 (1997): 203.
36. Angela Gugliotta, "How, When, and for Whom Was Smoke A Problem?" in *Devastation and Renewal: An Environmental History of Pittsburgh and*

Its Region, ed. Joel A. Tarr (Pittsburgh, PA: University of Pittsburgh Press, 2003), 118.

37. Gugliotta, "How, When, and for Whom," 121–23.
38. Sherrie R. Mershon and Joel A. Tarr, "Strategies for Clean Air, 1940–1960," in Tarr, *Devastation and Renewal*, 146.
39. "Mission, History and Vision of Magee-Womens Hospital of UPMC," accessed Mar. 10, 2017, http://www.upmc.com/locations/hospitals/magee/about-us/Pages/mission.aspx.
40. "Growing Green," *Environmental Initiatives of Magee-Womens Hospital of UPMC*, accessed Mar. 10, 2017, http://www.upmc.com/locations/hospitals/magee/about-us/environmental-initiatives/Pages/growing-green.aspx.
41. Christine O'Toole, "Magee: On the Green Frontier," *Annual Report of the Magee-Womens Research Institute and Foundation*, Winter 2013, 9–10.
42. Judith Focareta, RN, MEd, coordinator of environmental health initiatives, Magee-Womens Hospital of UPMC, author interview, Mar. 4, 2014.
43. Doug Oster, "Hospital Garden Yields Fresh Food," *Pittsburgh Post-Gazette*, Sept. 18, 2010.
44. Oster, "Hospital Garden."
45. Jennifer Silvis, "Leadership Walks, Talks, and Envisions a Healthier Future," *Healthcare Design*, Dec. 15, 2011, http://www.healthcaredesignmagazine.com/print/article/leadership-walks-talks-and-envisions-healthier-future.
46. Heinz Endowments and Magee-Womens Hospital of UPMC, Women's Health & the Environment: New Science, New Solutions Conference, Pittsburgh, PA, Apr. 20, 2007.
47. David Moffat, "Allegheny Riverfront Park, Pittsburgh Pennsylvania," *Places* 15, no. 1 (2002): 1, https://placesjournal.org/assets/legacy/pdfs/allegheny-riverfront-park-pittsburgh-pennsylvania.pdf.
48. Moffat, "Allegheny Riverfront Park," 1.
49. Michael Van Valkenburgh Associates, "Allegheny Riverfront Park Project," accessed Mar. 11, 2017, http://www.mvvainc.com/project.php?id=5; and Pittsburgh Cultural Trust, "Allegheny Riverfront Park," accessed Mar. 15, 2017, https://trustarts.org/pct_home/visual-arts/public-art/allegheny-riverfront-park/.
50. D. Gedge, "Roofspace: A Place For Brownfield Biodiversity?" *Ecos* 22, no. 3/4 (2002): 69–74.
51. The David Lawrence Convention Center was LEED Platinum building cer-

tified in 2012 by the Green Building Alliance. David Lawrence Convention Center, "Our Building: Building History," accessed Feb. 26, 2017, http://www.pittsburghcc.com/building/history.php.

52. The Green and Healthy Schools Academy has achieved partnerships with seventeen schools ranging from large public school districts to private faith-based and charter schools. See the list of partners at: *Green and Healthy Schools Academy*, accessed May 10, 2016, http://www.greenandhealthyschoolsacademy.org/schools.html.
53. Green Building Alliance, "Mission," *Green and Healthy Schools Academy*, accessed May 10, 2016, http://www.greenandhealthyschoolsacademy.org/vision--values.html.
54. Jenna Cramer, introduction of students from the Environmental Charter School, Green Building Alliance, Inspire Speaker Series, Apr. 10, 2014.
55. Gary J. Niels, "Welcome to Winchester Thurston School!," *Winchester Thurston*, accessed May 10, 2016, http://www.winchesterthurston.org/page.cfm?p=447.
56. Teresa DeFlitch, director of learning innovation, Winchester Thurston Academy, author interview, Aug. 3, 2012.
57. DeFlitch, author interview, Aug. 3, 2012.
58. Scott Weidensaul, *Living on the Wind: Across the Hemisphere with Migratory Birds* (New York: North Point Press, 1999), 365–70.
59. National Aviary, "Neighborhood Nestwatch," accessed Mar. 15, 2017, https://www.aviary.org/neighborhoodnestwatch.
60. Robert Mulvihill, ornithologist and program director for the Neighborhood Nestwatch program at the National Aviary, author interview, May 22, 2014. See also National Aviary, *Neighborhood Nestwatch*.

CHAPTER 4. TRANSFORMING THE ENERGY SYSTEM

Epigraph: Denis Hayes, *Rays of Hope: The Transition to a Post-Petroleum World* (New York: W. W. Norton, 1977), 156–57.

1. US Environmental Protection Agency, *Fact Sheet: Clean Power Plan Framework*, June 4, 2014, accessed Apr. 10, 2016, https://www.epa.gov/cleanpowerplan/fact-sheet-clean-power-plan-framework.

2. David Howard Davis, *Energy Politics*, 4th ed. (New York: St. Martin's, 1993), 26.
3. Kevin Hillstrom, *U.S. Environmental Policy and Politics: A Documentary History* (Washington, DC: CQ Press, 2010), 91.
4. Hillstrom, *U.S. Environmental Policy and Politics*, 162–64.
5. Peter Fox-Penner, *Electric Utility Restructuring: A Guide to the Competitive Era* (Vienna, VA: Public Utilities Reports, 1998), 11–12.
6. Leonard S. Hyman, Andrew S. Hyman, and Robert C. Hyman, *America's Electric Utilities: Past, Present, and Future*, 8th ed. (Vienna, VA: Public Utilities Reports, 2005), 147–48.
7. Harrison Brown, "Energy in Our Future," *Annual Review of Energy* 1 (1976): 11.
8. Richard M. Nixon, "Address to the Nation about National Energy Policy," Nov. 25, 1973, Public Papers of the Presidents, no. 339, *American Presidency Project*, accessed Sept. 9, 2014, http://www.presidency.ucsb.edu/ws/?pid=4051.
9. Jimmy Carter, "Proposed Energy Policy," *PBS American Experience*, broadcast Mar. 18, 1977, accessed June 5, 2015, http://www.pbs.org/wgbh/americanexperience/features/primary-resources/carter-energy/.
10. US Energy Information Administration, *Annual Energy Outlook 2017*, no. AEO2017, Jan. 5, 2017, 23, accessed Mar. 26, 2017, https://www.eia.gov/outlooks/aeo/pdf/0383(2017).pdf.
11. Ronald Reagan, "Message to the Congress Transmitting the National Energy Policy Plan," July 17, 1981, Public Papers of the Presidents, *American Presidency Project*, accessed Sept. 9, 2014, http://www.presidency.ucsb.edu/ws/?pid=44096.
12. US Energy Information Administration, *Annual Energy Outlook 2017*, 17.
13. US Energy Information Administration, *Annual Energy Outlook 2017*, 9.
14. Ronald Reagan, "First Inaugural Address," Jan. 20, 1981, *American Presidency Project*, accessed July 15, 2014, http://www.presidency.ucsb.edu/ws/?pid=43130.
15. Ken Silverstein, "Enron, Ethics and Today's Corporate Values," *Forbes Business*, May 14, 2014, http://www.forbes.com/sites/kensilverstein/2013/05/14/enron-ethics-and-todays-corporate-values/.
16. Energy Information Administration, "Status of Electricity Restructuring by State," data as of Sept. 2010, accessed Feb. 16, 2017, http://www.eia.gov/electricity/policies/restructuring/restructure_elect.html.

17. North Carolina Clean Energy Center, *Database of State Incentives for Renewables and Efficiency*, accessed Apr. 15, 2016, http://www.dsireusa.org/.
18. US Environmental Protection Agency, *Inventory of U.S. Greenhouse Gas Emissions and Sinks: 1990–2014*, EPA report 430-R-16-002 (Washington, DC: US Environmental Protection Agency, 2016), https://www.epa.gov/sites/production/files/2016-04/documents/us-ghg-inventory-2016-main-text.pdf.
19. Hyman, Hyman, and Hyman, *America's Electric Utilities*, 194.
20. *British thermal unit* (Btu) is a common measure of work representing the amount of work required to raise one pound of water one degree Fahrenheit. It is commonly used as a unit of comparison among energy sources. For a list of "Btu content of common energy units," see US Energy Information Administration, "Energy Units and Calculators Explained," *Energy Explained*, accessed Jan. 30, 2017, http://www.eia.gov/Energyexplained/?page=about_energy_units.
21. Lawrence Livermore National Laboratory, "Estimated U.S. Energy Consumption in 2016: 97.3 Quads," *Energy Flow Charts: Charting the Complex Relationships among Energy, Water and Carbon*, accessed Mar. 26, 2017, https://flowcharts.llnl.gov/.
22. Molly A. Maupin, Joan F. Kenny, Susan S. Hutson, John K. Lovelace, Nancy L. Barber, and Kristin S. Linsey, *Estimated Use of Water in the United States in 2010*, US Geological Survey circular 1405 (Reston, VA: US Geological Survey, 2014), 8, 40, http://pubs.usgs.gov/circ/1405/pdf/circ1405.pdf.
23. Peter Hanlon, Robin Madel, Kai Olson-Sawyer, Kyle Rabin, and James Rose, *Food, Water and Energy: Know the Nexus* (New York: GRACE Communications Foundation, 2013), 8, 19, http://www.gracelinks.org/media/pdf/knowthenexus_final_051513.pdf.
24. United States Senate, Complete Opening Statement of Rick Perry, Nominee for US Secretary of Energy, US Senate Committee on Energy and Natural Resources, Jan. 19, 2017, 3, https://www.energy.senate.gov/public/index.cfm/2017/1/nomination-hearing-of-the-honorable-rick-perry-for-secretary-of-energy.
25. Leo Gerard, President of International Steelworkers Union, keynote address at Good Jobs/Green Jobs Conference, May 4–6, 2010, Washington, DC, personal notes of the author.
26. International Energy Agency, *Energy and Climate Change*, World Energy

Outlook Special Report, Executive Summary (Paris, France: International Energy Agency, 2015), 3–6, http://www.iea.org/publications/freepublications/publication/WEO2015SpecialReportonEnergyandClimateChange ExecutiveSummaryUKversionWEB.PDF.

27. Carlo Cottarelli, Antoinette M. Sayeh, and Masood Ahmed, eds., *Energy Subsidy Reform: Lessons and Implications* (Washington, DC: International Monetary Fund, 2013), 4, http://www.imf.org/external/np/pp/eng/2013/012813.pdf.
28. Barack Obama, "Remarks on Oil and Gas Subsidies," Mar. 29, 2012, accessed Sept. 23, 2014, http://www.whitehouse.gov/photos-and-video/video/2012/03/29/president-obama-speaks-repealing-subsidies-oil-companies#transcript.
29. Camila Stark, Jacquelyn Pless, Jeffrey Logan, Ella Zhou, and Douglas J. Arent, *Renewable Electricity: Insights for the Coming Decade* (Golden, CO: Joint Institute for Strategic Energy Analysis, 2015), http://www.nrel.gov/docs/fy15osti/63604.pdf.
30. Renee Lewis Kosnik, *The Oil and Gas Industry's Exclusions and Exemptions from Major Environmental Statutes*, (Durango, CO: Oil and Gas Accountability Project, 2007), https://www.earthworksaction.org/files/publications/PetroleumExemptions1c.pdf.
31. Jonathan L. Ramseur, *Oil Spills in U.S. Coastal Waters: Background and Governance* (Washington, DC: US Congressional Research Service, 2012), 7–9, https://www.fas.org/sgp/crs/misc/RL33705.pdf.
32. Amory B. Lovins, *Soft Energy Paths: Toward a Durable Peace* (Cambridge, MA: Ballinger, 1977); and Amory B. Lovins and the Rocky Mountain Institute, *Reinventing Fire: Bold Business Solutions for the New Energy Era* (White River Junction, VT: Chelsea Green, 2011); and Paul Hawken, Amory B. Lovins, and L. Hunter Lovins, *Natural Capitalism: The Next Industrial Revolution*, 10th ed. (Washington, DC: Earthscan, 2010).
33. David Waskow and Jennifer Morgan, "The Paris Agreement: Turning Point for a Climate Solution," *World Resources Institute*, Dec. 12, 2015, accessed Aug. 16, 2016, http://www.wri.org/blog/2015/12/paris-agreement-turning-point-climate-solution.
34. Green Chemistry Roundtable, Institute for Green Science, Carnegie Mellon University, convened by the author on Nov. 6, 2014, and Apr. 21, 2015.
35. Energy Information Administration, "Natural Gas Pipelines in the North-

east Region," accessed June 10, 2015, http://www.eia.gov/pub/oil_gas/natural_gas/analysis_publications/ngpipeline/northeast.html.

36. Fuel Cell Energy Inc., "Company Overview," accessed Nov. 4, 2014, http://www.fuelcellenergy.com/about-us/company-overview/.
37. Nora Goldstein, "Farm Digester Progress in Pennsylvania," *BioCycle* 54, no. 5 (2013): 26. http://www.biocycle.net/2013/05/13/farm-digester-progress-in-pennsylvania/.
38. "Intelligent Workplace: A 'Living' and 'Lived-In' Lab," *Innovation & Creativity*, accessed June 1, 2015, http://www.cmu.edu/homepage/innovation/2007/spring/intelligent-workplace.shtml.
39. Farrokh Rahimi and Sasan Mokhtari, "From ISO to DSO," *Public Utilities Fortnightly* (June 2014): 42–50.
40. Pieter Gagnon, Robert Margolis, Jennifer Melius, Caleb Phillips, and Ryan Elmore, *Rooftop Solar Photovoltaic Technical Potential in the United States: A Detailed Assessment*, TP-6A20-65298 (Denver, CO: National Renewable Energy Laboratory, 2016), 11–14, http://www.nrel.gov/docs/fy16osti/65298.pdf.
41. Anthony Lopez, Billy Roberts, Donna Heimiller, Nate Blair, and Gian Porro, *U.S. Renewable Energy Technical Potentials: A GIS Based Analysis*, NREL/TP-6A20 (Denver, CO: National Renewable Energy Laboratory, July 2012), 3–5, http://www.nrel.gov/docs/fy12osti/51946.pdf.
42. Hal Saville, Energy Project Consultant, EIS Solar, Pittsburgh, Pennsylvania, author interview, May 15, 2013.
43. Saville, author interview, May 15, 2013.
44. Green Chemistry Roundtable, Nov. 6, 2014, and Apr. 21, 2015.
45. National Renewable Energy Laboratory, *Army Net Zero: Energy Roadmap and Program Summary: Fiscal Year 2013* (Washington, DC: US Department of Defense, 2014), 3, http://www.nrel.gov/docs/fy14osti/60992.pdf.
46. Dennis Yablonsky, "Energy and Innovation: Fueling the Present, Future Power of Pittsburgh," *Pittsburgh Business Times*, Oct. 14, 2011, http://www.bizjournals.com/pittsburgh/print-edition/2011/10/14/energy-and-innovation-fueling-pittsburgh.html.
47. Ordinance supplementing the Pittsburgh Code, Title Six, Conduct, Article 1, "Regulated Rights and Actions," by adding Chapter 619 entitled "Toxic Trespass Resulting from Unconventional Natural Gas Drilling."
48. William Peduto, "The Next Pittsburgh," inaugural speech, Jan. 6, 2014, *Office*

of Mayor William Peduto, accessed July 10, 2016, http://www.pittsburghpa.gov/mayor/inauguration.

49. Peduto, "Next Pittsburgh."
50. Almono, "Hazelwood Flats Redevelopment Map," accessed June 12, 2015, http://almono.org/map/hazelwood-flats.
51. SolarEdge, EIS Millvale Public Library Dashboard, accessed Apr. 18, 2016, https://monitoringpublic.solaredge.com/solaredge-web/p/site/public?name=EIS%20Millvale%20Community%20Library.
52. Jason Coughlin, Jennifer Grove, Linda Irvine, Janet F. Jacobs, Sarah Johnson Phillips, Alexandra Sawyer, and Joseph Wiedman, *A Guide to Community Shared Solar: Utility, Private, and Nonprofit Project Development* (Denver, CO: National Renewable Energy Laboratory, 2012), 3, http://www.nrel.gov/docs/fy12osti/54570.pdf.
53. Solar United Western PA, *2014 Solar Tour Guide Book* (Pittsburgh: Solar United Western Pennsylvania, 2014), 11.
54. Michael Carnahan, general manager, Scalo Solar Solutions, Pittsburgh, PA, author interview, June 24, 2014.
55. Carnahan, author interview, June 24, 2014; see also *Scalo Solar Solutions*, www.scalosolar.com.
56. Carnahan, author interview, June 24, 2014.
57. Carnahan, author interview, June 24, 2014.
58. Jack Scalo, president, Scalo Solar Solutions, Pittsburgh, PA, author interview, June 24, 2014.
59. Carnahan, author interview, June 24, 2014.
60. Carnahan, author interview, June 24, 2014.
61. Solar Energy Industries Association, "Third-Party Solar Financing," accessed May 10, 2016, http://www.seia.org/policy/finance-tax/third-party-financing.
62. Carnahan, author interview, June 24, 2014.
63. Carnahan, author interview, June 24, 2014.
64. US Department of Energy, Office of Energy Efficiency and Renewable Energy, "Advantages and Challenges of Wind Energy," accessed July 18, 2016, http://energy.gov/eere/wind/advantages-and-challenges-wind-energy.
65. Union of Concerned Scientists Report, "Farming the Wind: Wind Power and Agriculture," accessed June 10, 2015, http://www.ucsusa.org/clean_energy/smart-energy-solutions/increase-renewables/farming-the-wind-wind-power.html#.Va5ukMZViko.

66. D. L. Elliott, C. G. Holladay, W. R. Barchet, H. P. Foote, and W. F. Sandusky, "Chapter 2: The National Wind Resource," *Wind Energy Resource Atlas of the United States*, http://rredc.nrel.gov/wind/pubs/atlas/chp2.html.
67. Charles J. Barnhart, Michael Dale, Adam R. Brandt, and Sally M. Benson, "The Energetic Implications of Curtailing versus Storing Solar- and Wind-Generated Electricity," *Energy and Environmental Science* 6 (2013): 2804, http://pubs.rsc.org/en/content/articlepdf/2013/ee/c3ee41973h.
68. National Wind Coordinating Collaborative, *Wind Turbine Interactions with Birds, Bats, and Their Habitats: A Summary of Research Results and Priority Questions* (Washington, DC: US Department of Energy, 2010), http://www1.eere.energy.gov/wind/pdfs/birds_and_bats_fact_sheet.pdf.
69. Tom Zeller Jr., "For Those Near, the Miserable Hum of Clean Energy," *New York Times*, Oct. 5, 2010, http://www.nytimes.com/2010/10/06/business/energy-environment/06noise.html?_r=0.
70. Anya Litvak, "WindStax Owner Not Shying Away from Wind," *Pittsburgh Business Times*, Jan. 11, 2013, http://www.bizjournals.com/pittsburgh/print-edition/2013/01/11/windstax-founder-not-shying-away-wind.html.
71. Justine Coyne, "WindStax Looking to Quadruple Space, Add Employees," *Pittsburgh Business Times*, Aug. 4, 2014, http://www.bizjournals.com/pittsburgh/blog/innovation/2014/08/windstax-looking-to-quadruple-space-add-employees.html?s=print.
72. Michael Wilson, "Attention, Mr. Mayor: We Had Windmills 400 Years Ago," *New York Times*, Aug. 20, 2008, http://www.nytimes.com/2008/08/21/nyregion/21windside.html?_r=0.
73. Ron Gdovic, president, WindStax Wind Power Systems, Pittsburgh, PA, author interview, Aug. 6, 2014.
74. WindStax Wind Power Systems, "Alternative Wind Turbine Manufacturer Breezes into New Plant in Pittsburgh's Hi[s]toric Strip District," May 7, 2013, WindStax Big News Archives, accessed Feb. 16, 2017, http://windstax.com/about/big-news/big-news-archives/.
75. Gdovic, author interview, Aug. 6, 2014.
76. Eric Schwartzel, "WindStax Has Windmills for a Variety of Unexpected Places," *Pittsburgh Post-Gazette*, June 9, 2013, http://www.post-gazette.com/business/businessnews/2013/06/09/WindStax-has-windmills-for-a-variety-of-unexpected-places/stories/201306090188.
77. Gdovic, author interview, Aug. 6, 2014.

78. Justine Coyne, "Could Pittsburgh Be a Hub for Clean Technology?" *Pittsburgh Business Times*, July 31, 2014, http://www.bizjournals.com/pittsburgh/blog/innovation/2014/07/could-pittsburgh-be-a-hub-for-clean-technology.html.
79. Gdovic, author interview, Aug. 6, 2014.

CHAPTER 5. RESTORING FERTILE GROUND

Epigraphs: Vandana Shiva, "Annadana: The Gift of Food," in *Spiritual Ecology: The Cry of the Earth*, ed. Llewellyn Vaughan-Lee (Port Reyes, CA: Golden Sufi Center, 2013), 103; Rachel Carson, *Silent Spring* (Boston: Houghton Mifflin, 1962), 127; A.V. Krebs, "Corporate Takeover of Agriculture," in Andrew Kimbrell, *Fatal Harvest: The Tragedy of Industrial Agriculture* (Washington, DC: Island Press, 2002), 308.

1. Lindsay P. Smith, Shu Wen Ng, and Barry M. Popkin, "Trends in US Home Food Preparation and Consumption: Analysis of National Nutrition and Time Use Studies from 1965–1966 to 2007–2008," *Nutrition Journal* 12 (2013), doi:10.1186/1475-2891-12-45.
2. US Department of Agriculture, "Profiling Food Consumption in America," in *Agriculture Fact Book 2001–2002* (Washington, DC: US Government Printing Office, 2003), 15–16, https://www.usda.gov/documents/usda-factbook-2001-2002.pdf.
3. Food and Agricultural Organization of the United Nations, *Livestock and Landscapes* (Rome, Italy: Food and Agricultural Organization, 2012), accessed July 18, 2016, http://www.fao.org/docrep/018/ar591e/ar591e.pdf.
4. United Nations Environment Programme, *Avoiding Future Famines: Strengthening the Ecological Foundation of Food Security through Sustainable Food Systems* (Nairobi, Kenya: United Nations Environment Programme, 2012), 5, https://mahb.stanford.edu/wp-content/uploads/2013/02/2012-UNEP-Avoiding-Famines-Food-Security-Report.pdf.
5. Tim Searchinger, Craig Hanson, Janet Ranganathan, Brian Lipinski, Richard Waite, Robert Winterbottom, Ayesha Dinshaw, and Ralph Heimlich, *Creating a Sustainable Food Future*, World Resources Report 2013–2014: Interim Solutions (Washington, DC: World Resources Institute, 2014), 12–14, http://www.unep.org/gpa/documents/publications/CreatingSustainableFoodFuture.pdf.

6. United Nations Environment Programme, *Avoiding Future Famines*, 3.
7. William M. Muir, "The Threats and Benefits of GM Fish," *EMBO Reports* 5, no. 7 (2004): 654.
8. Tom Vilsak, table 1, "Historical Highlights 2012 and Earlier Census Years," *2012 Census of Agriculture* (Washington, DC: US Department of Agriculture, May 2014), 7, accessed Mar. 26, 2017, https://www.agcensus.usda.gov/Publications/2012/Full_Report/Volume_1,_Chapter_1_US/usv1.pdf.
9. US Department of Agriculture, Economic Research Service Statistics, "Adoption of Genetically Engineered Crops in the U.S.," http://www.ers.usda.gov/data-products/adoption-of-genetically-engineered-crops-in-the-us.aspx.
10. Mississippi River/Gulf of Mexico Watershed Nutrient Task Force, *2015 Report to Congress* (Washington, DC: US Environmental Protection Agency, 2015), 1, 15, accessed Mar. 19, 2017, https://www.epa.gov/sites/production/files/2015-10/documents/htf_report_to_congress_final_-_10.1.15.pdf.
11. G. Tyler Miller, *Sustaining the Earth: An Integrated Approach*, 6th ed. (Pacific Grove, CA: Thompson Learning, 2004), 211–16.
12. Jorge Fernandez-Cornejo, Richard Nehring, Craig Osteen, Seth Wechsler, Andrew Martin, and Alex Vialou, *Pesticide Use in U.S. Agriculture: 21 Selected Crops, 1960–2008, Economic Information Bulletin Number 124* (Washington, DC: US Department of Agriculture, Economic Research Service, 2014), 59–60, https://www.ers.usda.gov/webdocs/publications/eib124/46734_eib124.pdf.
13. Kimbrell, *Fatal Harvest*, 71.
14. Achim Steiner, "Preface," *The Environmental Food Crisis: The Environment's Role in Averting Future Food Crises*, ed. Christian Nellemann, Monika MacDevette, Ton Manders, Bas Eickhout, Birger Svihus, Anne Gerdien Prins, and Bjørn P. Kaltenborn (Arendal, Norway: United Nations Environment Programme, 2009), 5.
15. US Department of Agriculture, *Farms, Land in Farms, and Livestock Operations: 2012 Summary*, Feb. 19, 2013, http://usda.mannlib.cornell.edu/usda/nass/FarmLandIn/2010s/2013/FarmLandIn-02-19-2013.pdf.
16. United States Energy Information Administration, "Natural Gas Explained: Where Our Natural Gas Comes From," map, "Lower 48 States Shale Plays," accessed Feb. 21, 2017, https://www.eia.gov/energy_in_brief/article/shale_in_the_united_states.cfm.
17. American Farmland Trust, *Farming on the Edge: Sprawling Development*

Threatens America's Best Farmland (Washington, DC: American Farmland Trust, 2002), 4–5, http://www.farmlandinfo.org/sites/default/files/Farming_on_the_Edge_2002_1.pdf.

18. Transcript of "Acidifying Waters Corrode Northwest Shellfish," *PBS Newshour* segment reported by Hari Srinivasan, posted on National Oceanic and Atmospheric Administration, "Ocean Acidification's Impact on Oysters and Other Shellfish," *Pacific Marine Environmental Laboratory Carbon Program*, accessed Feb. 4, 2015, http://www.pmel.noaa.gov/co2/story/Ocean+Acidification%27s+impact+on+oysters+and+other+shellfish.
19. Food and Agriculture Organization of the United Nations, "Agriculture's Greenhouse Gas Emissions on the Rise," Apr. 11, 2014, accessed June 12, 2015, http://www.fao.org/news/story/en/item/216137/icode/.
20. C. Ford Runge, "The Case against Biofuels: Probing Ethanol's Hidden Costs," *Yale Environment 360*, Mar. 11, 2010, accessed Jan. 2015, http://e360.yale.edu/feature/the_case_against_biofuels_probing_ethanols_hidden_costs.
21. Mark W. Rosegrant and Siwa Msangi, "Consensus and Contention in the Food-Versus-Fuel Debate," *Annual Review of Environment and Resources* 39 (2014): 271–94, doi:10.1146/annurev-environ-031813-132233.
22. US Environmental Protection Agency, "Bio-Based Products and Chemicals, Waste-to-Energy Scoping Analysis" (Washington, DC: US Environmental Protection Agency, 2015), 3–5, https://www.epa.gov/sites/production/files/2015-12/documents/bio-based_products_and_chemicals_waste-to-energy_scoping_analysis_04032015_508.pdf.
23. Mark Koba, "Not Even Severe Drought Can Stop Fracking," *CNBC News*, June 10, 2014, accessed June 30 2014, http://www.cnbc.com/id/101738612.
24. Susan S. Hutson, Nancy L. Barber, Joan F. Kenny, Kristin S. Linsey, Deborah S. Lumia, and Molly A. Maupin, "Estimated Use of Water in the United States in 2000," USGS Circular 1268, last revised Feb. 2005, accessed June 10, 2014, https://pubs.usgs.gov/circ/2004/circ1268/.
25. Peter H. Gleick, testimony before the Subcommittee on Water Resources and Environment of the Committee on Transportation and Infrastructure, United States Congress, Public Hearing "Water: Is It the 'Oil' of the 21st Century?" June 4, 2003, *Pacific Institute*, accessed Feb. 21, 2017, http://pacinst.org/publication/water-efficiency-testimony-2/.
26. John A. Harper, "The Marcellus Shale—An Old 'New' Gas Reservoir in Pennsylvania," *Pennsylvania Geology* 38, no. 1 (2008): 6, accessed Mar. 17,

2017, http://www.dcnr.state.pa.us/cs/groups/public/documents/document/dcnr_006811.pdf; American Farmland Trust, "Farming on the Edge: Sprawling Development Threatens America's Best Farmland, Pennsylvania," map, Mar. 5, 2003, accessed Mar. 17, 2017, http://162.242.222.244/resources/fote/pdfs/map_pennsylvania.pdf.

27. William Kandel, "Profile of Hired Farmworkers, A 2008 Update," Economic Research Report No. 60, Economic Research Service, U.S. Department of Agriculture, June 2008. Page 8. Accessed Mar. 15, 2017. https://www.ers.usda.gov/webdocs/publications/err60/12056_err60_1_.pdf?v=42255
28. Thomas Herz and Steven Zahniser, "Immigration and the Rural Workforce," *US Department of Agriculture Economic Research Service*, Feb. 3, 2017, accessed Apr. 21, 2017, https://www.ers.usda.gov/topics/in-the-news/immigration-and-the-rural-workforce/.
29. Arthur Grube, David Donaldson, Timothy Kiely, and La Wu, *Pesticides Industry Sales and Usage: 2006 and 2007 Market Estimates* (Washington, DC: US Environmental Protection Agency, 2011), https://www.epa.gov/sites/production/files/2015-10/documents/market_estimates2007.pdf.
30. Daniel D. Chiras, *Environmental Science: Creating a Sustainable Future*, 6th ed. (Sudbury, MA: Jones and Bartlett, 2004), 539.
31. US Department of Agriculture, "Sustainable Agriculture: Definitions and Terms. Related Terms," *Alternative Farming Systems Information Center*, accessed Mar. 15, 2017, https://www.nal.usda.gov/afsic/sustainable-agriculture-definitions-and-terms-related-terms.
32. B. P. Baker, C. M. Benbrook, E. Groth III, K. Lutz Benbrook, "Pesticide Residues in Conventional, Integrated Pest Management (IPM)-Grown and Organic Foods: Insights from Three US Data Sets," *Food Additives and Contaminants* 19, no. 5 (2002): 427–46.
33. Baker et al., "Pesticide Residues," 427–46.
34. Baker et al., "Pesticide Residues," 427–46.
35. US Department of Agriculture, "Organic Production: Documentation," accessed Mar. 15, 2017, https://www.ers.usda.gov/data-products/organic-production/documentation/.
36. US Department of Agriculture, "Organic Production: Documentation," accessed May 10, 2016, http://www.ers.usda.gov/data-products/organic-production/documentation.aspx.
37. David Pimentel, Paul Hepperly, James Hanson, David Douds, and Rita

Seidel, "Environmental, Energetic, and Economic Comparisons of Organic and Conventional Farming Systems," *Bioscience*. 55, no. 7 (2005): 573–82.

38. Rodale Institute, *The Farming Systems Trial: Celebrating 30 Years* (Kutztown, PA: Rodale Institute, 2011), 8, https://rodaleinstitute.org/assets/FSTbooklet FINAL.pdf
39. Pimental et al., "Environmental, Energetic, and Economic Comparisons."
40. Susan S. Lang, "Organic Farms Produce Same Yields as Conventional Farms," *Cornell Chronicle*, July 13, 2005, http://www.news.cornell.edu/stories/2005/07/organic-farms-produce-same-yields-conventional-farms.
41. Mike Amaranthus, "Soil Carbon: Diamond in the Rough," *ACRES: The Voice of Eco-Agriculture*, Oct. 2008, 4.
42. Haitao Liu, Jing Li, Xiao Li, Yanhai Zheng, Sufei Feng, Gaoming Jiang, "Mitigating Greenhouse Gas Emissions through Replacement of Chemical Fertilizer with Organic Manure in a Temperate Farmland," *Science Bulletin* 60, no. 6 (2015): 598–606.
43. Rachel Carson, "A New Chapter to Silent Spring," in *Lost Woods: The Discovered Writing of Rachel Carson*, ed. Linda J. Lear (Boston: Beacon Press, 1998), 218.
44. HR 1599, the Safe and Accurate Food Labeling Act of 2015, passed the House but was defeated in the Senate. Colin O'Neil, "Senate Sends Strong Message in Defeat of DARK Act," *AgMag*, accessed May 6, 2016, http://www.ewg.org/agmag/2016/03/senate-sends-strong-message-defeat-dark-act.
45. Carson, "New Chapter," 222.
46. David Holmgren, *Permaculture: Principles and Pathways Beyond Sustainability* (Hepburn, Australia: Holmgren Design Services, 2002), xix.
47. Holmgren, *Permaculture*, xxviii.
48. Holmgren, *Permaculture*, 10.
49. Dana Gunders, *Wasted: How America Is Losing Up to 40 Percent of Its Food from Farm to Fork to Landfill*, NRDC Issue Paper 12-06-B (Washington, DC: National Resources Defense Council, 2012), 6, http://www.nrdc.org/food/files/wasted-food-ip.pdf.
50. Gunders, *Wasted*, 12.
51. Jenny Gustavsson, Christel Cederberg, Ulf Sonesson, Robert van Otterdijk, Alexandre Meybeck, *Global Food Losses and Food Waste: Extent, Causes and Prevention* (Rome, Italy: Food and Agriculture Organization of the United Nations, 2011), 4–6, http://www.fao.org/docrep/014/mb060e/mb060e.pdf.

52. Kevin D. Hall, Juen Guo, Michael Dore, Carson C. Chow, "The Progressive Increase of Food Waste in America and Its Environmental Impact," *PLoS ONE* 4, no. 11 (2009): e7940, doi:10.1371/journal.pone.0007940.
53. Gunders, *Wasted*, 7.
54. L. Kantor, K. Lipton, A. Manchester, and V. Oliveira, "Estimating and Addressing America's Food Losses," *USDA Food Review*, Jan.–Apr. 1997, 2-12.
55. Culling is the removal of products based on quality or appearance criteria, including specifications for size, color, weight, blemish level, and Brix (a measure of sugar content). Gunders, *Wasted*, 8.
56. Gunders, *Wasted*, 9.
57. Gunders, *Wasted*, 6.
58. Gunders, *Wasted*, 33.
59. Gunders, *Wasted*, 5.
60. US Environmental Protection Agency, "Future of Climate Change," *Climate Change Science*, accessed May 10, 2016, https://www.epa.gov/climate-change-science/future-climate-change.
61. *The New Urban Agriculture Zoning Code: What Does It Mean to You?*, accessed Dec. 10, 2015, http://www.growpittsburgh.org/wp-content/uploads//UAgZoningCode-SimpleEdition.pdf.
62. Shelly Danko-Day, open space specialist, City of Pittsburgh Planning Department, author interview, July 24, 2014.
63. "Our Mission, Vision and Values," *Grow Pittsburgh*, accessed Feb. 21, 2017, http://www.growpittsburgh.org/about-us/our-mission-vision-and-values/.
64. South Side Slopes Neighborhood Association, "South Side Community Garden at Bandi Schaum Field," *South Side Slopes*, accessed July 14, 2015, http://www.southsideslopes.org/the-neighborhood/community-garden/.
65. Danko-Day, author interview, July 24, 2014.
66. US Department of Agriculture, "Organic Production Documentation."
67. *Kretschmann Family Organic Farm*, accessed Jan. 9, 2015, http://www.kretschmannfarm.com.
68. Don Kretschmann, owner, Kretschmann Family Organic Farm, author interview, Sept. 12, 2014.
69. Don Kretschmann, "Kretschmann Farm Apples," accessed Jan. 12, 2015, http://archive.kretschmannfarm.com/Newsletters/AppleStory.html.
70. Kretschmann, author interview, Sept. 12, 2014.
71. *Kretschmann Farm, LLC v. Twp. of New Sewickley*, 2016 Pa. Commw. LEXIS

33 (Pa. Commw. Ct. 2016), accessed May 4, 2016, http://www.landuselawinpa.com/court/042716_kretschmann_farm.shtml.

72. Kretschmann, author interview, Dec. 30, 2014.

73. Marie Fechik-Kirk contributed research and writing to this case study.

74. *Eat'n Park Hospitality Group, Inc.*, accessed May 18, 2016, http://www.enphospitality.com/.

75. Eat'n Park Hospitality Group, "Sustainability," accessed Jan. 9, 2015, http://www.eatnpark.com/content.aspx?pid=3&ppid=0.

76. Recombinant bovine somatotrophin, rBST, is a synthetic hormone injected into dairy cattle to stimulate milk production. "Monsanto Urges FDA to Stop 'Misleading' rBST-Free Labeling," *The Organic & Non-GMO Report*, May 2007, accessed May 6, 2016, http://www.non-gmoreport.com/articles/may07/misleading_rBST-free_labeling.php.

77. Eat'n Park Hospitality Group, "Sustainability Initiatives."

78. Nick Camody served on the board of directors of the Rachel Carson Homestead Association from 2008 to 2012, and we had many personal conversations on this topic during my tenure there as executive director from 2006 to 2010. The Rachel Carson Legacy Challenge issued in 2007 as part of the Centennial Celebration stated:

> May 27, 2007, marked the centennial of author and ecologist Rachel Carson's birth in Springdale, Pennsylvania. The year-long centennial celebration included the Rachel Carson Legacy Challenge to individuals, businesses, organizations, institutions and government to make permanent, measurable changes in behavior and policies that promote Rachel Carson's environmental ethic, which is: To live in harmony with nature; To preserve and learn from natural places; To minimize the impact of man-made chemicals on natural systems of the world; and To consider the implications of human actions on the global web of life.
>
> The objective of this challenge is to create a platform to demonstrate how coordinated commitment to environmentally sustainable practices can make a real and tangible difference in the health, quality of life, environment, and economic viability of local, regional and global communities. As part of the web of life, people who adopt this challenge will build conditions for a more sustainable, healthy world.

Patricia DeMarco, executive director of the Rachel Carson Homestead Association, "Legacy Challenge," *Rachel Carson Homestead*, accessed Mar.

15, 2017, http://rachelcarsonhomestead.org/rachel-carson-legacy/legacy-challenge/.

79. For the Parkhurst Dining Services commitments, see "Rachel Carson Legacy Challenge Commitments," *Rachel Carson Homestead*, accessed Jan. 9, 2015, http://rachelcarsonhomestead.org/rachel-carson-legacy/legacy-challenge-commitments/.
80. Eat'n Park Hospitality Group, "Sustainability Initiatives."
81. Eat'n Park Hospitality Group, "Sustainability Initiatives."
82. James Moore, director of sourcing and sustainability, Eat'n Park Hospitality Group, author interview, Nov. 10, 2014.
83. Moore, author interview, Nov. 10, 2014.
84. Kevin O'Connell, senior vice-president for marketing, Eat'n Park Hospitality Group, author interview, Nov. 10, 2014.
85. Eat'n Park Hospitality Group, "Sustainability Initiatives."
86. Moore, author interview, Nov. 10, 2014.

CHAPTER 6. PREVENTING POLLUTION

Epigraphs: Rachel Carson, *Silent Spring* (New York: Houghton Mifflin, 1962), 12–13; Carson, *Silent Spring*, 15; Janine M. Benyus, *Biomimicry: Innovation Inspired by Nature* (New York: HarperCollins, 1997), 97.

1. For information about the Allegheny River Stewardship Project, led by principal investigator Conrad Daniel Volz, see University of Pittsburgh Graduate School of Public Health, "CHEC Projects," *Center for Healthy Environments and Communities*, accessed Apr. 28, 2016, http://www.chec.pitt.edu/projects.html.
2. Conrad Daniel Volz, principal investigator, Allegheny River Stewardship Project, author interview, Mar. 22, 2008.
3. Lara Huyler, "Exploratory Data Report/Estrogenicity of Shad Species," Allegheny River Stewardship Project, University of Pittsburgh Graduate School of Public Health, Center for Healthy Environments and Communities, Apr. 2, 2009, http://www.chec.pitt.edu/documents/ExploratoryDataRep-Huyler.pdf.
4. Conrad D. Volz and C. Christen, "Occupational Medicine Forum," *Journal of Occupational and Environmental Medicine* 49, no. 1 (2007): 104–5.

5. Katelyn Polantz, "Pittsburgh Rivers, Waters Full of Toxins," *Pitt News*, Mar. 21, 2007, accessed Feb. 17, 2017, http://pittnews.com/article/27461/archives/pittsburghs-rivers-water-full-of-toxins/.
6. United States Department of Health and Human Services, Centers for Disease Control and Prevention, *Fourth National Report on Human Exposure to Environmental Chemicals, 2009* (Atlanta, GA: Department of Health and Human Services, Centers for Disease Control and Prevention, Dec. 2009), http://www.cdc.gov/exposurereport/pdf/fourthreport.pdf.
7. National Institute of Environmental Health Sciences, "Bisphenol A (BPA)," accessed Mar. 17, 2017, https://www.niehs.nih.gov/health/topics/agents/sya-bpa/.
8. Centers for Disease Control and Prevention, "Factsheet: Bisphenol A," *National Biomonitoring Program*, accessed Mar. 23, 2107, https://www.cdc.gov/biomonitoring/bisphenola_factsheet.html.
9. Cynthia J. Hines, Matthew V. Jackson, James A. Deddens, John C. Clark, Xiaoyun Ye, Annette L. Christianson, Juliana W. Meadows, and Antonia M. Calafat, "Urinary Bisphenol A (BPA) Concentrations among Workers in Industries that Manufacture and Use BPA in the USA," *Annals of Work Exposure and Health* 61, no. 2 (2017): 164-82, doi:10.1093/annweh/wxw021.
10. Sonya Lunder, David Andrews, and Jane Houlihan, "BPA Coats Cash Register Receipts," *EWG Research*, July 27, 2010, accessed Mar. 24, 2017, http://www.ewg.org/research/bpa-in-store-receipts.
11. Carson, *Silent Spring*, 190.
12. The United States has an age-standardized cancer rate of 318 per 100,000. World Cancer Research Fund International, "Data for Cancer Frequency by Country," accessed June 2, 2015, http://www.wcrf.org/int/cancer-facts-figures/data-cancer-frequency-country.
13. Linda Birnbaum, "When Chemicals Disrupt," keynote address, Rachel Carson Legacy Conference, Rachel Carson Homestead Association, Pittsburgh, PA, Sept. 27, 2009.
14. Carson, *Silent Spring*, 13.
15. US Environmental Protection Agency, "TRI-Listed Chemicals," *Toxic Release Inventory Program*, accessed June 1, 2015, http://www2.epa.gov/toxics-release-inventory-tri-program/tri-listed-chemicals.
16. Amanda Porter and Michelle Corrigan, "Toxic Substances Control Act: Past, Present and Future," *Stinson Leonard Street*, June 28, 2016, accessed Mar. 23,

2017, https://www.stinson.com/Resources/Alerts/2016_Alerts/Toxic_Substances_Control_Act__Past,_Present_and_Future.aspx.

17. Jennifer Beth Sass and Aaron Colangelo, "European Union Bans Atrazine, while the United States Negotiates Continued Use," *International Journal of Occupational and Environmental Health* 12, no. 3 (2006): 260–67, http://dx.doi.org/10.1179/oeh.2006.12.3.260.
18. International Agency for Research on Cancer, *IARC Monographs Volume 112: Evaluation of Five Organophosphate Insecticides and Herbicides* (Lyons, France: World Health Organization, 2015), http://www.iarc.fr/en/media-centre/iarcnews/pdf/MonographVolume112.pdf.
19. Arthur Grube, David Donaldson, Timothy Kiely, and La Wu, "Table 3.6. Most Commonly Used Conventional Pesticide Active Ingredients, Agricultural Market Sector, 2007, 2005, 2003, and 2001 Estimates," and "Table 3.7. Most Commonly Used Conventional Pesticide Active Ingredients, Home and Market Sector, 2007, 2005, 2003, and 2001 Estimates," in *Pesticides Industry Sales and Usage: 2006 and 2007 Market Estimates* (Washington, DC: United States Environmental Protection Agency, 2011), https://www.epa.gov/sites/production/files/2015-10/documents/market_estimates2007.pdf.
20. Tyrone B. Hayes, Vicky Khoury, Anne Narayan, Mariam Nazir, Andrew Park, Travis Brown, Lillian Adame, Elton Chan, Daniel Buchholz, Theresa Stueve, and Sherri Gallipeau, "Atrazine Induces Complete Feminization and Chemical Castration in Male African Clawed Frogs (*Xenopus laevis*)," *PNAS* 107, no. 10 (2010): 4612–17.
21. Arlene Blum, Simona A. Balan, Martin Scheringer, Xenia Trier, Gretta Goldenman, Ian T. Cousins, Miriam Diamond, Tony Fletcher, Christopher Higgins, Avery E. Lindeman, Graham Peaslee, Pim de Voogt, Zhanyun Wang, and Roland Weber, "The Madrid Statement on Poly- and Perfluoroalkyl Substances (PFASs)," *Environmental Health Perspectives* 123, no. 5 (2015): A107–11, doi:10.1289/ehp.1509934.
22. Callie Lyons, *Stain-Resistant, Nonstick, Waterproof, and Lethal: The Hidden Dangers of C8* (Santa Barbara, CA: Praeger, 2007).
23. Brian Bienkowski, "Scientists Call For Limits on Stain- and Waterproofing Chemicals," *Environmental Health News*, May 1, 2015, accessed May 1, 2015, http://www.environmentalhealthnews.org/ehs/news/2015/may/fluorinated-chemicals-madrid-statement-science-health.
24. Mae Wu, Dylan Atchley, Linda Greer, Sarah Janssen, Daniel Rosenberg, and

Jennifer Sass, *Dosed without Prescription: Preventing Pharmaceutical Contamination of Our Nation's Drinking Water*, National Resources Defense Council, Dec. 2009, accessed June 1, 2015, http://docs.nrdc.org/health/files/hea_10012001a.pdf.

25. Janet M. Carter, Gregory C. Delzer, James A. Kingsbury, and Jessica A. Hopple, *Concentration Data for Anthropogenic Organic Compounds in Ground Water, Surface Water, and Finished Water of Selected Community Water Systems in the United States, 2002–05*, US Geological Survey Data Series 268 (Washington, DC: United States Geological Survey, 2007), http://pubs.usgs.gov/ds/2007/268/.
26. Åke Bergman, Jerrold J. Heindel, Susan Jobling, Karen A. Kidd, and R. Thomas Zoeller, eds., *State of the Science of Endocrine Disrupting Chemicals 2012* (Paris, France: United Nations Environment Programme and the World Health Organization, 2013).
27. United States Environmental Protection Agency, Code of Federal Regulations Subpart M–National Emission Standard for Asbestos, 40 CFR 763 Subpart I, accessed July 20, 2016, https://www.gpo.gov/fdsys/pkg/CFR-2011-title40-vol31/pdf/CFR-2011-title40-vol31-part763-subpartI.pdf.
28. US Environmental Protection Agency, "The Frank R. Lautenberg Chemical Safety for the 21st Century Act," accessed July 18, 2016, https://www.epa.gov/assessing-and-managing-chemicals-under-tsca/frank-r-lautenberg-chemical-safety-21st-century-act.
29. US Environmental Protection Agency, "Frank R. Lautenberg Chemical Safety."
30. Kenneth Cook, president of The Environmental Working Group, Testimony on S.697 before the Senate Committee on Environment and Public Works, Mar. 18, 2015, accessed May 3, 2015, http://www.ewg.org/testimony-official-correspondence/testimony-s697-senate-committee-environment-and-public-works.
31. Congressional Research Service, Summary: H.R. 4432—113th Congress (2013–2014), Safe and Accurate Food Labeling Act of 2014, accessed Mar. 25, 2015, www.congress.gov/bill/113th-congress/house-bill/4432. This law contains a provision preventing states from labeling foods containing genetically modified organisms, reserving the process to the federal Food and Drug Administration.
32. Naomi Oreskes and Erik M. Conway, *Merchants of Doubt: How a Handful of*

Scientists Obscured the Truth on Issues from Tobacco Smoke to Global Warming (New York: Bloomsbury, 2010), 160.

33. Elizabeth Meacham, assistant professor, Ursuline College, personal communication with the author, Mar. 20, 2015.
34. Oreskes and Conway, *Merchants of Doubt*, 236.
35. REACH is the European Regulation on Registration, Evaluation, Authorisation and Restrictions of Chemicals, entered into force in the European Union on June 1, 2007. European Chemicals Agency, "Understanding REACH," accessed Mar. 23, 2017, https://echa.europa.eu/regulations/reach/understanding-reach.
36. European Environmental Bureau, "EU Policy: The REACH Regulation" (Brussels: European Environmental Bureau, 2006), accessed Oct. 4, 2016, http://www.eeb.org/index.cfm/activities/industry-health/chemicals/reach/.
37. European Environmental Bureau, "EU Policy: REACH," accessed Oct. 4, 2016, http://www.eeb.org/index.cfm/activities/industry-health/chemicals/reach/.
38. Rachel Carson, "On the Pollution of the Environment," in *Lost Woods: The Discovered Writing of Rachel Carson*, ed. Linda J. Lear (Boston: Beacon, 1998), 243.
39. Terrence J. Collins, "Review of the Twenty-Three Year Evolution of the First University Course in Green Chemistry: Teaching Future Leaders How to Create Sustainable Societies," *Journal of Cleaner Production* 140 (2017): 93–110.
40. Terry Collins, "The Importance of Sustainability Ethics, Toxicity and Ecotoxicity in Chemical Education and Research," *Green Chemistry* 5 (2003): G51–52.
41. US Environmental Protection Agency, "Basics of Green Chemistry," accessed Mar. 19, 2015, http://www2.epa.gov/green-chemistry/basics-green-chemistry.
42. T. T. Schug, R. Abagyan, B. Blumberg, T. J. Collins, D. Crews, P. L. DeFur, S. M. D. T. M. Edwards, A. C. Gore, L. J. Guillette, T. Hayes, J. J. Heindel, A. Moores, H. B. Patisaul, T. L. Tal, K. A. Thayer, L. N. Vandenberg, J. C. Warner, C. S. Watson, F. S. Saal, R. T. Z. K. P. O'Brien, and J. P. Myers, "Designing Endocrine Disruption Out of the Next Generation of Chemicals," *Green Chemistry* 15 (2013): 181–98.
43. United States Department of the Interior, United States Geologic Survey, *USGS Circular 1292: Volatile Organic Compounds in the Nation's Ground*

Water and Drinking-Water Supply Wells (Washington, DC: US Department of the Interior), accessed Apr. 19, 2015, http://water.usgs.gov/nawqa/vocs/national_assessment/report.

44. Nathaniel R. Warner, Cidney A. Christie, Robert B. Jackson, and Avner Vengosh, "Impacts of Shale Gas Wastewater Disposal on Water Quality in Western Pennsylvania," *Environmental Science and Technology* 47, no. 20 (2013): 11849–57.
45. Brian Bienkowski, "Hormone-Mimickers Widespread in Great Lakes Region Wastewater, Waterways and Fish," *Environmental Health News*, Mar. 23, 2015, accessed Mar. 25, 2015, http://www.environmentalhealthnews.org/ehs/news/2015/mar/great-lakes-water-chemicals-fish-health/document_view.
46. Terrence J. Collins, Teresa Heinz Professor of Green Chemistry, Carnegie Mellon University, author interview, Apr. 21, 2015.
47. Terrence J. Collins and Chip Walter, "Little Green Molecules," *Scientific American* 294, no. 3 (2006): 84.
48. Collins and Walter, "Little Green Molecules," 85.
49. Collins and Walter, "Little Green Molecules," 82.
50. Collins and Walter, "Little Green Molecules," 87.
51. Collins, author interview, Apr. 21, 2015.
52. Matthew R. Mills, Karla Arias-Salazar, Alice Baynes, Longzhu Q. Shen, John Churchley, Nicola Beresford, Chakicherla Gayathri, Roberto G. Gil, Rakesh Kanda, Susan Jobling, and Terrence J. Collins, "Removal of Ecotoxicity of 17α-ethinylestradiol Using TAML/Peroxide Water Treatment," *Scientific Reports* 5, no. 10511 (2015), doi:10.1038/srep10511.
53. GreenOX Catalysts, "Technology and Markets," accessed Apr. 19, 2015, http://www.greenoxcatalysts.com.
54. Colin Horowitz, chief technology officer, GreenOX Catalysts, author interview, Apr. 27, 2015.
55. Collins, author interview, Apr. 27, 2015.
56. Eric Beckman, director, Mascaro Center for Sustainable Innovation, University of Pittsburgh, "Green Chemistry Roundtable I. Eliminating the Institutional Barriers," records of Patricia DeMarco, executive director, Rachel Carson Homestead Association, Pittsburgh, PA, May 7, 2009.
57. US Environmental Protection Agency, *Toxics Release Inventory (TRI) Program*, accessed Apr. 13, 2017, https://www.epa.gov/toxics-release-inventory-tri-program.

58. Christianna Briddell, "Eric Beckman: Integrating Green," *Green Chemistry: The Nexus Blog*, American Chemical Society, May 21, 2014, accessed Mar. 19, 2017, https://communities.acs.org/community/science/sustainability/green-chemistry-nexus-blog/blog/2014/05/21/eric-beckman-integrating-green.
59. Beckman, "Green Chemistry Roundtable I."
60. *Mascaro Center for Sustainable Innovation*, accessed Feb. 17, 2017, http://www.engineering.pitt.edu/MCSI/.
61. Michael J. Buckley and Eric J. Beckman, "Adhesive Use in Oral and Maxillofacial Surgery," *Oral and Maxillofacial Surgery Clinics of North America* 22, no. 1 (2010): 195–99, doi:10.1016/j.coms.2009.10.008.
62. Linda Wang, "Venturing Out: Scientists Transition into Entrepreneurship by Taking Risks and Following Their Passions," *Chemical and Engineering News*, Apr. 19, 2012, 50–52, http://cen.acs.org/articles/90/i16/Venturing.html.
63. *Cohera Medical, Inc.*, accessed Apr. 20, 2015, http://www.coheramedical.com/.
64. Joseph P. Hunstad, Joseph Michaels, A. Jay Burns, Sheri Slezak, W. Grant Stevens, Dottie M. Clower, and J. Peter Rubin, "A Prospective, Randomized, Multicenter Trial Assessing a Novel Lysine-Derived Urethane Adhesive in a Large Flap Surgical Procedure without Drains," *Aesthetic Plastic Surgery* 39, no. 4 (2015): 616–24, doi:10.1007/s00266-015-0498-4.
65. *Cohera Medical, Inc.*, accessed Apr. 20, 2015, http://www.coheramedical.com/.
66. Beckman, "Green Chemistry Roundtable I."
67. Madasyn Czebiniak, "Lalit Chordia of Thar Energy Focuses on the Green," *Pittsburgh Post-Gazette Powersource*, June 11, 2014, http://powersource.post-gazette.com/powersource/consumers-powersource/2014/06/10/Lalit-Chordia-of-Thar-Energy-focuses-on-the-green/stories/201406040196.
68. Lalit Chordia, president, Thar Technologies, Inc., Pittsburgh, PA, author interview, Mar. 12, 2015.
69. Madasyn Czebiniak, "Lalit Chordia of Thar Energy Focuses on the Green," *Pittsburgh Post-Gazette Powersource*, June 11, 2014, http://powersource.post-gazette.com/powersource/consumers-powersource/2014/06/10/Lalit-Chordia-of-Thar-Energy-focuses-on-the-green/stories/201406040196.
70. Tom Imerito, "Supercritical Man," *Science Spectrum*, accessed May 1, 2015, http://science-spectrum.com/supercritical-man-2/.
71. Chordia, author interview, Mar. 12, 2015.
72. Chordia, author interview, Mar. 12, 2015.

73. Chordia, author interview, Mar. 12, 2015.
74. Environmental Working Group, *Skin Deep Cosmetics Database*, accessed Feb. 20, 2017, www.ewg.org/skindeep/.
75. Women for a Healthy Environment, "Who We Are," accessed Feb. 20, 2017, http://www.womenforahealthyenvironment.org/who-we-are.html.
76. Teresa Heinz, "Women's Health and the Environment: New Science, New Solutions," conference proceedings, accessed June 1, 2015, http://www.womens healthpittsburgh.org/.
77. Michelle Naccarati-Chapkis, executive director, Women for a Healthy Environment, author interview, Mar. 12, 2015.
78. Kim T. Gordon, "Target Moms' Buying Power," *Entrepreneur*, June 1, 2009, http://www.entrepreneur.com/article/201976.
79. Arlene Blum, "Trading Health for Small Conveniences?" *MomsRising.org*, May 1, 2015, accessed Feb. 20, 2017, https://www.momsrising.org/blog/trading-health-for-small-conveniences.
80. *Safer Chemicals, Healthy Families*, accessed Apr. 19, 2015, http://saferchemicals.org/.
81. Laura Coleman-Lochner and Andrew Martin, "Wal-Mart Asks Its Suppliers to Stop Using Eight Chemicals," *Bloomberg*, July 20, 2016, accessed Mar. 19, 2017, https://www.bloomberg.com/news/articles/2016-07-20/wal-mart-asks-suppliers-to-remove-eight-chemicals-from-products.
82. Andy Ingrejas, "What Senators Aren't Getting about State Chemical Policies," *Safer Chemicals, Healthy Families*, Mar. 27, 2015, accessed Apr. 19, 2015, http://saferchemicals.org/2015/03/27/what-senators-arent-getting-about-state-chemical-policies/.
83. Safer Chemicals, Healthy Families, "Mind the Store," accessed Mar. 19, 2017, http://saferchemicals.org/mind-the-store/.
84. Safer Chemicals, Healthy Families, "The Hazardous 100+ List of Chemicals of High Concern," last updated July 17, 2013, accessed May 1, 2015, http://saferchemicals.org/wp-content/uploads/2014/05/mindthestore.org-full-list-toxic-chemicals2.pdf.
85. Naccarati-Chapkis, author interview, Mar. 12, 2015.
86. Naccarati-Chapkis, author interview, Mar. 12, 2015.
87. Naccarati-Chapkis, author interview, Mar. 12, 2015.
88. Naccarati-Chapkis, author interview, Mar. 12, 2015.

CHAPTER 7. MOBILIZING AWARENESS TO ACTION

Epigraphs: Wangari Maathai, *Unbowed: A Memoir* (New York: Alfred A. Knopf, 2006), 125; Andrew Slaughter and Hope Comden, extracts from "Visions for the Future" student assignment for GEOL 1055: Science, Ethics and Public Policy, taught by Patricia M. DeMarco, University of Pittsburgh, 2013, used with permission; Rachel Carson, "Biological Sciences," in *Lost Woods: The Discovered Writing of Rachel Carson*, ed. Linda J. Lear (Boston: Beacon Press, 1998), 167.

1. Donella Meadows, Jorgen Randers, Dennis Meadows, *Limits to Growth: The 30-Year Update* (White River Junction, VT: Chelsea Green, 2004), 273.
2. Lauren Feldman, Teresa A. Myers, Jay D. Hmielowski, and Anthony Leiserowitz, "The Mutual Reinforcement of Media Selectivity and Effects: Testing the Reinforcing Spirals Framework in the Context of Global Warming," *Journal of Communication* 64, no. 4 (2014): 590–611.
3. Anthony Leiserowitz, Edward Maibach, Connie Roser-Renouf, Geoff Feinberg, and Peter Howe, *Global Warming's Six Americas, September 2012* (New Haven, CT: Yale Project on Climate Change Communication, 2013), http://environment.yale.edu/climate-communication-OFF/files/Six-Americas-September-2012.pdf.
4. "Media 'Echo Chambers' and Climate Change," *Yale Program on Climate Change Communication*, Sept. 29, 2014, accessed Feb. 25, 2017, http://climatecommunication.yale.edu/publications/media-echo-chambers-and-climate-change/.
5. "Media 'Echo Chambers.'"
6. "Media 'Echo Chambers.'"
7. Lydia Saad, "Global Warming Concern at Three-Decade High in US," *Gallup*, Mar. 14, 2017, accessed Mar. 20, 2017, http://www.gallup.com/poll/206030/global-warming-concern-three-decade-high.aspx.
8. National Science Board, "Chapter 7. Science and Technology: Public Attitudes and Understanding," *Science and Engineering Indicators 2014*, accessed July 15, 2014, https://www.nsf.gov/statistics/seind14/index.cfm/chapter-7/c7h.htm.
9. Art Swift, "Americans Again Pick Environment over Economic Growth,"

Gallup, Mar. 20, 2017, accessed Mar. 22, 2017, http://www.gallup.com/poll/168017/americans-again-pick-environment-economic-growth.aspx.

10. Swift, "Americans."
11. Yvonne Campos, Campos Research Strategy, "Qualitative Research Report: Climate Change/Behavior Change," study completed for Patricia DeMarco, funded by the Pittsburgh Foundation W. Clyde and Ida Mae Thurman Fund, July 9, 2014.
12. Campos, "Qualitative Research Report," July 9, 2014.
13. Hart Research Associates, survey, "Pennsylvania Attitudes about EPA Carbon Regulations and Climate Change," June 3, 2014, https://web.archive.org/web/20150324141647/http://www.lcv.org/issues/polling/pa-poll-on-carbon-rules-and.pdf.
14. Anthony Leiserowitz, Edward Maibach, Connie Roser-Renouf, Matthew Cutler, and Seth Rosenthal, *Trump Voters and Global Warming* (New Haven, CT: Yale Program on Climate Change Communication), accessed Mar. 19, 2017, http://climatecommunication.yale.edu/wp-content/uploads/2017/02/Trump-Voters-and-Global-Warming.pdf.
15. Patricia DeMarco, Institute for Green Science, Carnegie Mellon University, "Roundtable on a New Utility Paradigm," audio recording file, Nov. 6, 2014.
16. DeMarco, "Roundtable on a New Utility Paradigm," audio recording file, Apr. 21, 2014.
17. Allegheny Conference on Community Development, "What We're Working On: Energy," accessed Apr. 8, 2015, https://web.archive.org/web/20150408032026/http://www.alleghenyconference.org/InitiativeEnergy.php.
18. Jeanne E. Moorman, Lara I. Akinbami, Cathy M. Bailey, Hatice S. Zahran, Michael E. King, Carol A. Johnson, and Xiang Liu, *National Surveillance of Asthma: United States, 2001–2010*, DHHS publication no. (PHS) 2013–1419, Vital Health Statistics series 3, no. 35 (Washington, DC: US Government Printing Office, 2012), 11, accessed Mar. 24, 2017, https://www.cdc.gov/nchs/data/series/sr_03/sr03_035.pdf.
19. American Lung Association, *State of the Air 2016* (Chicago, IL: American Lung Association., 2016), 8, http://www.lung.org/assets/documents/healthy-air/state-of-the-air/sota-2016-full.pdf.
20. US Environmental Protection Agency, "Healthier Americans," *Mercury and*

Air Toxics Standards, accessed Feb. 24, 2017, https://www.epa.gov/mats/healthier-americans#methylmercury.

21. Breathe Project, "About the Breathe Project," accessed June 5, 2015, http://breatheproject.org/about/; Heinz Endowments Report on the Breathe Project, accessed June 5, 2015, http://www.heinz.org/Interior.aspx?id=323.
22. David Templeton, "How Polluted Is the Pittsburgh Region? CMU Professor Helps Shed Light on Air Quality," *Pittsburgh Post-Gazette*, Feb. 11, 2015, accessed Mar. 25, 2017, http://www.post-gazette.com/news/environment/2015/02/11/Carnegie-Mellon-University-professor-makes-new-pollution-maps-public/stories/201502110023.
23. Breathe Project, "Making It Count: Breathe Project Coalition Partners Are Taking Actions that Add Up to Cleaner Air," accessed Aug. 16, 2016, http://breatheproject.org/act/making-it-count/.
24. Allegheny County Health Department, "Air Quality: Monitored Data," accessed Aug. 16, 2016, http://www.achd.net/air/monitor-data.html.
25. Pennsylvania Public Utility Commission, "Act 129 Information," http://www.puc.state.pa.us/filing_resources/issues_laws_regulations/act_129_information.aspx.
26. Jeaneen Zappa, executive director, Conservation Consultants, Inc., author interview, Feb. 18, 2015.
27. Zappa, author interview, Feb. 18, 2015.
28. Zappa, author interview, Feb. 18, 2015.
29. Zappa, author interview, Feb. 18, 2015.
30. Zappa, author interview, Feb. 18, 2015.
31. Zappa, author interview, Feb. 18, 2015.
32. Pittsburgh Green Workplace Challenge, *2014–2015 Green Workplace Challenge Results* (Pittsburgh, PA: Sustainable Pittsburgh, 2016), 3, http://gwcpgh.org/images/2014-2015PittsburghGWCResults.pdf.
33. Pittsburgh Green Workplace Challenge, "Total Impact GWC 1.0 - 3.0: Cumulative GWC Results," accessed Mar. 20, 2017, http://gwcpgh.org/impact/total-impact.
34. Matthew Mehalik, program manager, Sustainable Pittsburgh, author interview, Feb. 23, 2015.
35. Pittsburgh Green Workplace Challenge, *2014–2015 Competition Guidebook:*

Version (12.23.2014) (Pittsburgh: Sustainable Pittsburgh, 2014), http://gwcpgh.org/images/GWC_Guide_V_3_15_111014.pdf.

36. Mehalik, author interview, Feb. 23, 2015.
37. Andrew Butcher, Juan Cristiani, Bethany Davidson, Duy Ngo, Ann M. Rosenthal, Nathan Wildfire, and Laura Zamarripa, *Vacant to Vibrant: A Guide for Revitalizing Vacant Lots* (Pittsburgh, PA: Carnegie Mellon University, H. John Heinz III, School of Public Policy and Management, 2006).
38. GTECH, "Who We Are," accessed Feb. 21, 2017, https://gtechstrategies.org/who-we-are/.
39. Andrew Butcher, president, GTECH Strategies, Pittsburgh, PA, author interview, May 21, 2014.
40. Sara Innamorato, "Grease for Good," *GTECH News*, Dec. 17, 2013, accessed Feb. 21, 2017, https://gtechstrategies.org/pittsburgh-oil-recycling/.
41. GTECH, "What We Do," accessed Sept. 8, 2015, https://web.archive.org/web/20150908035340/https://gtechstrategies.org/what-we-do/our-approach/.
42. GTECH Strategies, *ReClaim*, accessed Mar. 31, 2015, https://gtechstrategies.org/what-we-do/reclaim/.
43. Butcher, author interview, May 21, 2014.
44. Andrew Butcher, president, GTECH Strategies, interview by Marie Fechik-Kirk, Mar. 3, 2015.
45. Butcher, interview by Fechik-Kirk, Mar. 3, 2015.
46. Butcher, interview by Fechik-Kirk, Mar. 3, 2015.
47. Rhonda Sears, GTECH Community Ambassador, Pittsburgh, PA, interview by Marie Fechik-Kirk, Dec. 15, 2014.
48. Sears, interview by Fechik-Kirk, Dec. 15, 2014.
49. Sears, interview by Fechik-Kirk, Dec. 15, 2014.
50. GTECH Strategies, "Get to Know . . . Rhonda Sears," Aug. 22, 2014, accessed Feb. 21, 2017, https://gtechstrategies.org/reenergizepittsburgh_rhonda/.
51. Sears, interview by Fechik-Kirk, Dec. 15, 2014.
52. Butcher, author interview, May 21, 2014.
53. Anna Archer, "ReClaim Ambassadors Graduate!," *GTECH*, Oct. 8, 2015, accessed Mar. 21, 2017, https://gtechstrategies.org/reclaim-ambassadors-graduate/.
54. Frank Borman quoted in "CT Borman Assigned Higher Post; 3 Astronauts Report to Nation," *Palm Beach Post*, Jan. 10, 1969, 27.

CHAPTER 8. THE NEW ECONOMY

Epigraphs: Pope Francis, "Infected by a Culture of Waste," General Papal Address Commemorating World Environment Day, June 5, 2013, accessed Aug. 12, 2016, http://www.ewtn.com/library/PAPALDOC/f1genaud11.htm; Tim Jackson, *Prosperity without Growth: Economics for a Finite Planet* (London: Earthscan, 2009), 187; Herman E. Daly, *Beyond Growth: The Economics of Sustainable Development* (Boston: Beacon Press, 1996), 201.

1. Jenna R. Jambeck, Roland Geyer, Chris Wilcox, Theodore Siegler, Miriam Perryman, Anthony Andrady, Ramani Namayan, and Kara Lavender Law, "Plastic Waste Inputs from Land into the Ocean," *Science* 347, no. 6223 (2015): 768–71.
2. ExxonMobil, "Guiding Principles," accessed Apr. 28, 2016, http://corporate.exxonmobil.com/en/company/about-us/guiding-principles/our-guiding-principles.
3. ExxonMobil, "Guiding Principles."
4. ExxonMobil spent $11.98 million on lobbying in 2015 in the United States. Figure acquired from US Senate website, http://www.senate.gov, accessed Apr. 28, 2016, search for "ExxonMobil" as registrant name.
5. World Commission on Environment and Development, *United Nations Report of the World Commission on Environment and Development: Our Common Future* (Oslo, Sweden: United Nations, 1987), 7.
6. World Commission on Environment and Development, *United Nations Report of the World Commission on Environment and Development: Our Common Future* (Oslo, Sweden: United Nations, Mar. 20, 1987), 18.
7. Paul Hawken, Amory B. Lovins, and L. Hunter Lovins, *Natural Capitalism: The Next Industrial Revolution*, 10th ed. (Washington, DC: Earthscan, 2010), 5.
8. Hawken, Lovins, and Lovins, *Natural Capitalism*, xi.
9. American Sustainable Business Council, "Member Businesses," accessed Mar. 24, 2017, http://asbcouncil.org/membership/member-businesses#.WNhUZWTyvUQ.
10. Andrew W. Savitz and Karl Weber, *The Triple Bottom Line* (San Francisco, CA: Jossey-Bass, 2006), 70–72.
11. Office of Mayor William Peduto, "Mayor William Peduto and 100 Resilient

Cities—Pioneered by the Rockefeller Foundation Host 'Kickoff' Workshop Marking Beginning of City's Participation in Innovative Global Urban Resilience Initiative," press release, June 5, 2015, http://pittsburghpa.gov/mayor/release?id=4583; p4 Performance Measures Project, *p4 Performance Measures* (Pittsburgh, PA: City of Pittsburgh, 2016), 3, http://p4pittsburgh.org/media/W1siZiIsIjIwMTYvMTAvMTEvbmd5a3FsYWZnX3A0X1BlcmZvcm1hbmNlX01lYXN1cmVzX2ZpbmFsLnBkZiJdXQ/p4-Performance_Measures-final.pdf.

12. Hawken, Lovins, and Lovins, *Natural Capitalism*, 4, 285–308.
13. American Petroleum Institute, *An American Energy Revolution: Energy in Charts 2016* (Washington, DC: American Petroleum Institute, 2016), 1–93, http://www.api.org/~/media/Files/Policy/American-Energy/american-energy-in-charts-online.pdf.
14. Devin N. Perkins, Marie-Noel Brune Drisse, Tapiwa Nxele, and Peter D. Sly, "E-Waste: A Global Hazard," *Annals of Global Health* 80, no. 4 (2014): 286–95.
15. Energy Information Administration, *Annual Energy Outlook Report, May 2015*, accessed June 15, 2015, http://www.eia.gov/forecasts/aeo/.
16. Gar Alperovitz, "The Political-Economic Foundations of a Sustainable System," in *State of the World 2014: Governing for Sustainability*, ed. Worldwatch Institute (Washington, DC: Worldwatch Institute, 2014), 195.
17. Alperovitz, "Political-Economic Foundations," 196.
18. Alperovitz, "Political-Economic Foundations," 195.
19. Gar Alperovitz, "The New-Economy Movement," *Nation*, May 25, 2011, http://www.thenation.com/article/160949/new-economy-movement.
20. Alperovitz, "New-Economy Movement."
21. Alperovitz, "Political-Economic Foundations."
22. Joe Uehlein, founding president and executive director, Labor Network for Sustainability, author interview, Aug. 22, 2015.
23. *American Sustainable Business Council*, "Overview," accessed May 21, 2015, http://asbcouncil.org/about-us#.VWIDntpViko. In the spirit of full disclosure: the author is a member of the ASBC as DeMarco and Associates, LLC, since 2013.
24. Marie Fechik-Kirk contributed to writing this case study.
25. Alcoa Inc. has produced a sustainability report since 2002. The 2016 Alcoa Corporation report, which will be published in 2017, will reflect the sustain-

ability performance of current operations—bauxite, alumina, aluminum, cast products, rolled products, and energy. Past sustainability reports can be found at www.alcoa.com/sustainability. It is important to note for comparison purposes that these reports include information and data on businesses that are no longer part of today's Alcoa following the company's separation in 2016 and divestitures through the years. Alcoa Corporation, *Alcoa Sustainability*, accessed Mar. 26, 2017, https://www.alcoa.com/sustainability/en/pdf/Alcoa-Sustainability.pdf.

26. William O'Rourke, executive director of the Beard Institute in the Polumbo-Donahue School of Business at Duquesne University, author interview, Jan. 29, 2015.
27. O'Rourke, author interview, Jan. 29, 2015.
28. Roy Harvey, "CEO Statement on Sustainability," Alcoa Corporation, *Alcoa Sustainability*.
29. O'Rourke, author interview, Jan. 29, 2015.
30. William McDonough and Michael Braungart, *Cradle to Cradle: Remaking the Way We Make Things* (New York: North Point, 2002), 81–83.
31. Alcoa, "Who We Are," accessed Feb. 21, 2017, http://www.alcoa.com/global/en/who-we-are/default.asp.
32. William O'Rourke, interview by Marie Fechik-Kirk, Apr. 9, 2015.
33. Cradle to Cradle Products Innovation Institute, "Get Cradle to Cradle Certified," accessed Feb. 21, 2017, http://www.c2ccertified.org/get-certified/product-certification.
34. Alcoa Corporation, *2015 Alcoa Sustainability Report*, accessed Feb. 21, 2017, http://www.alcoa.com/sustainability/en/pdf/archive/corporate/2015-Sustainability-Report.pdf; Alcoa Corporation, *2013 Alcoa Sustainability at a Glance*, accessed Feb. 21, 2017, http://www.alcoa.com/sustainability/en/pdf/archive/corporate/2013_Sustainability_Highlights_Report.pdf.
35. United States Environmental Protection Agency, table 3.1, "Electronic Waste Management in the United States, Approach 1," EPA 530-R-08-009 (Washington, DC: US Environmental Protection Agency, July 2008), 25, http://www.epa.gov/osw/conserve/materials/ecycling/docs/app-1.pdf.
36. United States Environmental Protection Agency, "Municipal Solid Waste Generation, Recycling and Disposal in the United States; Tables and Figures for 2012," (Washington, DC: United States Environmental Protection Agen-

cy, 2014), tables 12–14, http://www.epa.gov/epawaste/nonhaz/municipal/pubs/2012_msw_dat_tbls.pdf.

37. Lina Younes, "Don't Let That Used Phone Go to Waste," *Our Planet, Our Home*, Nov. 14, 2013, accessed May 11, 2015, https://blog.epa.gov/blog/2013/11/dont-let-that-used-phone-go-to-waste/.
38. Stefan Schwarzer, Andréa De Bono, Gregory Guiliani, Stéphane Kluser, and Pascal Peduzzi, *E-Waste, The Hidden Side of IT Equipment's Manufacturing and Use*, Environment Alert Bulletin 5 (Geneva, Switzerland: United Nations Environment Programme, 2005).
39. Eric Williams, "Environmental Impacts in the Production of Personal Computers," in *Computers and the Environment: Understanding and Managing Their Impact*, ed. Ruediger Kuehr and Eric Williams (Dordrecht, Netherlands: Kluwer Academic Publishers, 2003), 41–72.
40. Ned Eldridge, president of eLoop, LLC, presentation at Champions 4 Sustainability Workshop, Sustainable Pittsburgh, Pittsburgh, PA, Mar. 14, 2008, author notes.
41. Ned Eldridge, president, eLoop, LLC, author interview, Apr. 15, 2015.
42. Eldridge, author interview, Apr. 15, 2015.
43. Pennsylvania Department of Environmental Protection, *The Covered Device Recycling Act (Act 108) of 2010 (CDRA), A General Overview*, accessed Mar. 25, 2017, http://www.elibrary.dep.state.pa.us/dsweb/Get/Document-109540/2520-FS-DEP4324.pdf.
44. Pennsylvania Department of Environmental Protection, "Manufacturer Requirements," accessed Mar. 26, 2017, http://www.dep.pa.gov/Business/Land/Waste/Recycling/Electronics/Pages/Manufacturer-Requirements.aspx.
45. Eldridge, author interview, Apr. 15, 2015.
46. CBS News, "Electronics Wasteland," *60 Minutes*, Nov. 9, 2008, accessed May 15, 2015, http://www.cbs.com/shows/60_minutes/video/923894885/the-electronic-wasteland/.
47. Eldridge, author interview, Apr. 15, 2015. Basel Action Network (BAN) is the world's only organization focused on confronting the global environmental injustice and economic inefficiency of toxic trade (toxic wastes, products, and technologies) and its devastating impacts. Basel Action Network, "About Us," accessed May 15, 2015, http://www.ban.org/about/.

48. Eldridge, author interview, Apr. 15, 2015.
49. eLoop, LLC, "Environmental Stewardship," accessed Mar. 26, 2017, http://eloopllc.com/about-us/environental-stewardship/.
50. PennFuture is a nonprofit organization that has been operating in Pennsylvania since 1998 as an environmental advocacy group. See the website at http://www.pennfuture.org/.
51. Eldridge, author interview, Apr. 15, 2015.
52. Eldridge, author interview, Apr. 15, 2015.
53. LaKeisha Wolf, executive director of Ujamaa Collective and manager of Ujamaa Boutique, author interview, May 16, 2015.
54. *Ujamaa Collective*, accessed May 15, 2015, www.ujamaacollective.org.
55. Patrick Doyle, "Pittsburgh's Hill District, Reimagined," *Pittsburgh Magazine*, Jan. 22, 2015, accessed May 15, 2015, http://www.pittsburghmagazine.com/Pittsburgh-Magazine/February-2015/Pittsburghs-Hill-District-Reimagined/.
56. Wolf, author interview, May 16, 2015.
57. Ujamaa Collective, "The Marketplace," accessed May 17, 2015, http://www.ujamaacollective.org/the-marketplace/.
58. Wolf, author interview, May 16, 2015.
59. ABAFASI, "About," accessed Mar. 19, 2017, https://abafasiabafasi.wordpress.com/about/.
60. "Gran'ma's Hands" performed at the Freedom from Fracking Concert, May 16, 2015, Mr. Small's Theater, Pittsburgh, PA.
61. The Cooperation over Competition program expands the collective through membership to support the work of Ujamaa. "Cooperation over Competition by Ujamaa Collective," *Global Switchboard*, May 8, 2015, accessed Feb. 21, 2017, http://theglobalswitchboard.org/2015/05/cop-ujamaa-collective/.

CHAPTER 9. BUILDING A RESILIENT CITY

Epigraphs: Office of Mayor William Peduto, "Mayor William Peduto's Public Schedule Dec. 2, 2015," press release, Dec. 2, 2015, accessed June 12, 2016, http://pittsburghpa.gov/mayor/release?id=5430; International Living Future Institute, *Living Building Challenge 3.0*, accessed Apr. 24, 2017, https://access.living-future.org/lbc/about; Rachel Carson, "The Real World around Us," in

Lost Woods: The Discovered Writing of Rachel Carson, ed. Linda J. Lear (Boston: Beacon Press, 1998), 163.

1. Jeremy Rifkin, *The Third Industrial Revolution: How Lateral Power Is Transforming Energy, the Economy, and the World* (New York: Palgrave Macmillan, 2013), 36–38.
2. Ian Morris, "Imagining a World after Fossil Fuels," *Forbes*, Mar. 23, 2017, accessed Mar. 23, 2017, https://www.forbes.com/sites/stratfor/2017/03/23/imagining-a-world-after-fossil-fuels/#29e0c29e6d6f.
3. City of Pittsburgh, Department of Innovation and Performance, *Resilient Pittsburgh: Final Assessment Report* (Pittsburgh, PA: City of Pittsburgh, 2016), http://apps.pittsburghpa.gov/cis/PRA2016_Final_version.pdf.
4. William Peduto, "100 Resilient Cities Pittsburgh Workshop #4–Mayor's Address and Closing," John Heinz Regional History Museum, Pittsburgh, PA, June 5, 2015, accessed Feb. 26, 2017, https://www.youtube.com/watch?v=wgRR9NIV8tM&feature=youtu.be&list=PL2L3ZTwe78kHmYAAbXf_PAjP3OCWVOL9D.
5. Office of Mayor Peduto, "Mayor Peduto's Public Schedule December 2, 2015," accessed Mar. 26, 2017, http://pittsburghpa.gov/mayor/release?id=5430.
6. *Pittsburgh 2030 District Progress Report* (Pittsburgh, PA: Green Building Alliance, 2014), http://www.2030districts.org/sites/default/files/atoms/files/Pittsburgh%202030%20District%202013%20Progress%20Report.pdf.
7. Sam Kusic, "Yablonsky: Wolf's Severance Tax Plan Will Hurt Development," *Pittsburgh Business Times*, Mar. 27, 2015, accessed July 18, 2016, http://www.bizjournals.com/pittsburgh/blog/energy/2015/03/yablonsky-wolfs-severance-tax-plan-will-hurt.html.
8. Allegheny Conference on Community Development, New Center of American Energy, accessed June 5, 2015, http://alleghenyconference.org/InitiativeEnergy.php.
9. David Conti, "Pollution Potential of Beaver County Plant Sizable," *Trib Live*, June 19, 2014, accessed July 22, 2016, http://triblive.com/business/headlines/6304823-74/plant-shell-credits.
10. Ken Silverstein, "Creative Disruption," *Public Utilities Fortnightly*, Apr. 2015, accessed Feb. 26, 2017, https://www.fortnightly.com/fortnightly/2015/04/creative-disruption.
11. City of Pittsburgh, *Resilient Pittsburgh*.
12. Sally Seppanen and Wendy Gualtieri, *The Millennial Generation: Research*

Review (Washington, DC: National Chamber Foundation, 2012), accessed June 5, 2015, https://www.uschamberfoundation.org/sites/default/files/article/foundation/MillennialGeneration.pdf.

13. Housing Authority of the City of Pittsburgh, *Larimer Vision-to-Action Plan*, Phase I, (Pittsburgh, PA: Housing Authority of the City of Pittsburgh, 2012), 4, accessed June 4, 2015, https://larimerplan.files.wordpress.com/2013/08/larimer-vision-to-action-plan-report_phase-i.pdf.
14. Larimer Neighborhood Collaborative, *Larimer Community Plan* (Pittsburgh: Larimer Neighborhood Collaborative, 2008), accessed June 2, 2015, https://larimerplan.files.wordpress.com/2008/11/larimer_communityplan.pdf.
15. Housing Authority of the City of Pittsburgh, *Larimer Vision-to-Action Plan*, 4.
16. Fred Brown, associate director for program development, Kingsley Association, author interview, July 27, 2014.
17. Housing Authority of the City of Pittsburgh, *Larimer Vision-to-Action Plan*, 1.
18. Brown, author interview, July 27, 2014.
19. Diana Nelson Jones, "Pittsburgh Lands $30 Million Grant to Rebuild Larimer," *Pittsburgh Post-Gazette*, June 30, 2014, http://www.post-gazette.com/local/city/2014/06/30/City-lands-30-million-grant-to-rebuild-Larimer/stories/201406300171.
20. Brown, author interview, July 27, 2014.
21. EcoDistricts, "Our Theory of Change," accessed June 1, 2015, http://ecodistricts.org/about/.
22. Grant Ervin, manager for sustainability and resilience for the Office of the Mayor, City of Pittsburgh, author interview, Apr. 28, 2015.
23. Kathy Riciutto, "Incubator Success Story: Bringing Up Uptown in Pittsburgh, PA," *EcoDistricts*, Feb. 23, 2015, accessed June 1, 2015, http://ecodistricts.org/incubator-success-story-bringing-up-uptown-in-pittsburgh-pa.
24. Ervin, author interview, Apr. 28, 2015.
25. Fourth Economy Consulting, "Uptown Community Vision Plan," version 1.0, Apr. 17, 2009, accessed Mar. 23, 2017, http://www.uptownpartners.org/wp-content/uploads/2010/03/Uptown-Vision-one-document.pdf.
26. Riciutto, "Incubator Success Story."
27. Riciutto, "Incubator Success Story."
28. Ed Blazina, "How Would a Bus Rapid Transit Project Benefit the Pittsburgh

Region?" *Pittsburgh Post-Gazette*, Mar. 9, 2017, accessed Mar. 23, 2017, http://www.post-gazette.com/local/city/2017/03/09/How-would-a-Bus-Rapid-Transit-project-Uptown-Oakland-benefit-the-Pittsburgh-region/stories/201703100086.

29. Riciutto, "Incubator Success Story."
30. William Peduto, mayor, City of Pittsburgh, author interview, May 21, 2015.
31. Peduto, author interview, May 21, 2015.
32. Peduto, author interview, May 21, 2015.
33. Peduto, author interview, May 21, 2015.
34. City of Pittsburgh Office of Sustainability, "Green Buildings in Pittsburgh," accessed Nov. 5, 2014, http://www.pittsburghpa.gov/green/buildings.htm.
35. Richard Piacentini, executive director, Phipps Conservatory and Botanical Gardens, author interview, Mar. 19, 2014.
36. Mary Adam Thomas, *Building in Bloom* (Portland, OR: Ecotone Publishing, 2013), 2.
37. International Living Future Institute, "Living Building Challenge, Petals," accessed Mar. 23, 2017, https://living-future.org/lbc/.
38. Thomas, *Building in Bloom*, 16–17.
39. Phipps Conservatory and Botanical Gardens, "About," accessed June 12, 2015, http://phipps.conservatory.org/about-phipps/index.aspx.
40. Thomas, *Building in Bloom*, 19.
41. Thomas, *Building in Bloom*, 12–13.
42. Piacentini, author interview, Mar. 19, 2014.
43. Thomas, *Building in Bloom*, 17.
44. Thomas, *Building in Bloom*, 37.
45. Phipps Conservatory and Botanical Gardens, "Center for Sustainable Landscapes," accessed Feb. 26, 2017, https://phipps.conservatory.org/green-innovation/at-phipps/center-for-sustainable-landscapes-greenest-building-in-the-world.
46. Thomas, *Building in Bloom*, 28.
47. Vivian Loftness, professor of architecture, Carnegie Mellon University, author interview, May 15, 2014.
48. Piacentini, author interview, Mar. 19, 2014.
49. Piacentini, author interview, Mar. 19, 2014.
50. Phipps Conservatory and Botanical Garden, "Green Heart of Pittsburgh," accessed May 12, 2015, https://phipps.conservatory.org/green-innovation/.

51. Diana Nelson Jones, "Pittsburgh Recognized as Biophilic City," *Pittsburgh Post-Gazette*, Sept. 17, 2016, accessed Mar. 23, 2017, http://www.post-gazette.com/local/city/2016/09/17/Pittsburgh-s-cap-gets-another-feather-with-induction-into-Biophilic-Cities-network-an-effort-to-integrate-nature-into-daily-city-life/stories/201609170031.
52. Jones, "Pittsburgh Recognized."
53. Edward O. Wilson, *Half-Earth: Our Planet's Fight for Life* (New York: Liveright, 2016).

CONCLUSION: THE POWER OF JOINED VOICES

Epigraphs: Rachel Carson, *Silent Spring* (Boston: Houghton Mifflin Company, 1962), 6; Ban Ki-Moon, "Secretary-General's Remarks at Workshop on the Moral Dimensions of Climate Change and Sustainable Development 'Protect the Earth, Dignify Humanity' [As Delivered]," Vatican City, Apr. 28, 2015, http://www.un.org/sg/statements/index.asp?nid=8584.

1. William Peduto, "The Next Pittsburgh," *Pittsburgh Post-Gazette*, January 6, 2014, accessed June 5, 2015, http://www.post-gazette.com/local/city/2014/01/06/Trasncript-Bill-Peduto-s-inaugural-speech/stories/201401060154, used with permission.

SELECTED BIBLIOGRAPHY

The field of sustainability merges across many disciplines. The foundations rest on understanding the physical and biological laws that drive the interconnected systems of the earth. This collection of references and resources offers some of the historical foundations as well as some of the important cutting-edge thinking. This is not intended to be a comprehensive reference list, but may provoke exploration among the many intersecting disciplines of this dynamic field.

Alcamo, Josef. *Avoiding Future Famines: Strengthening the Ecological Foundation of Food Security through Sustainable Food Systems*. Nairobi: United Nations Environment Programme, 2012. http://www.unep.org/publications/ebooks/avoidingfamines/portals/19/UNEP_Food_Security_Report.pdf.

Alperovitz, Gar. "The New-Economy Movement." *Nation*. May 25, 2011. http://www.thenation.com/article/160949/new-economy-movement.

Beckman, Jayson, Allison Borchers, and Carol A. Jones. *Agriculture's Supply and Demand for Energy and Energy Products*. Economic Information Bulletin 112. Washington, DC: US Department of Agriculture, Economic Research Service, May 2013. https://www.ers.usda.gov/webdocs/publications/eib112/37427_eib112.pdf.

Benyus, Janine M. *Biomimicry: Innovation Inspired by Nature.* New York: Harper Perennial, 1997.

Bullard, Robert D., ed. *The Quest for Environmental Justice: Human Rights and the Politics of Pollution.* San Francisco: Sierra Club Books, 2005.

Bergman, Åke, Jerrold J. Heindel, Susan Jobling, Karen A. Kidd, and R. Thomas Zoeller, eds. *State of the Science of Endocrine Disrupting Chemicals, 2012.* Geneva, Switzerland: United Nations Environment Programme and the World Health Organization, 2013.

Brundtland, Gro Harlem. *United Nations Report of the World Commission on Environment and Development: Our Common Future.* Oslo: United Nations, 1987.

Cardinale, Bradley J., J. Emmett Duffy, Andrew Gonzalez, David U. Hooper, Charles Perrings, Patrick Venail, Anita Narwani, Georgina M. Mace, David Tilman, David A. Wardle, Ann P. Kinzig, Gretchen C. Daily, Michel Loreau, James B. Grace, Anne Larigauderie, Diane S. Srivastava, and Shahid Naeem. "Biodiversity Loss and Its Impact on Humanity." *Nature* 486, no. 7401 (2012): 59–67. doi: 10.1038/nature1114807.

Carson, Rachel. *Guarding Our Wildlife Resources.* Conservation in Action Series, no. 5. Washington, DC: US Department of the Interior, Fish and Wildlife Service, 1943.

Carson, Rachel. *The Sea around Us,* illustrated commemorative ed. New York: Oxford University Press, 2003.

Carson, Rachel. *Silent Spring.* Boston: Houghton Mifflin, 1962.

Carson, Rachel. *The Sense of Wonder.* New York: HarperCollins, 1998.

Carson, Rachel. Testimony. Hearings before the US Senate Subcommittee on Reorganization and International Organizations of the Committee on Government Operations, "Interagency Coordination in Environmental Hazards (Pesticides)." June 4, 1963, 88th Congress, 1st sess. Washington, DC: Government Printing Office, 1964.

"Case Studies." *The Natural Step.* Accessed July 16, 2016. http://www.thenaturalstep.org/cases/.

Chiras, Daniel D. *Environmental Science: Creating a Sustainable Future.* 6th ed. Sudbury, MA: Jones and Bartlett, 2004.

City of Pittsburgh, Department of Innovation and Performance. *Resilient Pittsburgh.* Pittsburgh, PA: City of Pittsburgh, June 2016. http://apps.pittsburghpa.gov/cis/PRA2016_Final_version.pdf.

Colborn, Theo, Dianne Dumanoski, and John Peterson Myers. *Our Stolen Future: Are We Threatening Our Fertility, Intelligence, and Survival? A Scientific Detective Story*. New York: Penguin, 1997.

Collins, Terrence J. "The Importance of Sustainability Ethics, Toxicity and Ecotoxicity in Chemical Education and Research." *Green Chemistry* 5, no. 4 (2003): G51–G52.

Collins, Terrence J. "Review of the Twenty-Three Year Evolution of the First University Course in Green Chemistry: Teaching Future Leaders How to Create Sustainable Societies." *Journal of Cleaner Production* 140 (2015): 93–110.

Costanza, Robert, Ralph d'Arge, Rudolf de Groot, Stephen Farber, Monica Grasso, Bruce Hannon, Karin Limburg, Shahid Naeem, Robert V. O'Neill, Jose Paruelo, Robert G. Raskin, Paul Sutton, and Marjan van den Belt. "The Value of the World's Ecosystem Services and Natural Capital," *Nature* 387, no. 6630 (1997): 253–60.

Crocker, David A., and Toby Linden., eds. *Ethics of Consumption—The Good Life, Justice, and Global Stewardship*. Lanham, MD: Rowman and Littlefield, 1998.

Dernbach, John C. *Acting as if Tomorrow Matters: Accelerating the Transition to Sustainability*. Washington, DC: Environmental Law Institute, 2012.

Dunbar-Ortiz, Roxanne. *An Indigenous Peoples' History of the United States*. Boston: Beacon Press, 2014.

Eddy, John. *A New Sun: The Solar Results from Skylab*. Edited by Rein Ise. Washington, DC: National Aeronautics and Space Administration, 1979. http://history.nasa.gov/SP-402/sp402.htm.

Ehrenfeld, John R. *Sustainability by Design: A Subversive Strategy for Transforming Our Consumer Culture*. New Haven, CT: Yale University Press, 2008.

Ehrlich, Paul R. *Human Natures: Genes, Cultures, and the Human Prospect*. Washington, DC: Island Press, 2000.

Ehrlich, Paul R., and John P. Holdren. "Impact of Population Growth." *Science* 171, no. 3977 (1971): 1212–17. doi: 10.1126/science.171.3977.1212.

Elliott, D. L., C. G. Holladay, W. R. Barchet, H. P. Foote, and W. F. Sandusky. "The National Wind Resource." In *Wind Energy Resource Atlas of the United States*. Washington, DC: US Department of Energy, 1986. http://rredc.nrel.gov/wind/pubs/atlas/chp2.html.

Energy Information Administration. *Annual Energy Outlook 2015*. Washington, DC: US Energy Information Administration, 2015. http://www.eia.gov/outlooks/archive/aeo15/pdf/0383(2015).pdf.

Esty, Daniel C., and Andrew S. Winston. *Green to Gold: How Smart Companies Use Environmental Strategy to Innovate, Create Value, and Build Competitive Advantage.* New Haven, CT: Yale University Press, 2006.

Feldman, Lauren, Teresa A. Myers, Jay D. Hmielowski, and Anthony Leiserowitz. "The Mutual Reinforcement Of Media Selectivity and Effects: Testing the Reinforcing Spirals Framework in the Context of Global Warming." *Journal of Communication* 64, no. 4 (2014): 590–611.

Fox-Penner, Peter S. *Electric Utility Restructuring: A Guide to the Competitive Era.* Vienna, VA: Public Utilities Reports, 1998.

Francis. *Laudato Si': On Care for Our Common Home.* Vatican City, Italy: Libreria Editrice Vaticana, 2015. Accessed January 30, 2017, http://w2.vatican.va/content/francesco/en/encyclicals/documents/papa-francesco_20150524_enciclica-laudato-si.html

Gagnon, Pieter, Robert Margolis, Jennifer Melius, Caleb Phillips, and Ryan Elmore. *Rooftop Solar Photovoltaic Technical Potential in the United States: A Detailed Assessment.* Technical Report TP-6A20-65298. Golden, CO: National Renewable Energy Laboratory, 2016. http://www.nrel.gov/docs/fy16osti/65298.pdf.

Gardiner, Stephen Mark. "A Perfect Moral Storm: Climate Change, Intergenerational Ethics, and the Problem of Moral Corruption." *Environmental Values* 15 (2006): 397–413.

Gunders, Dana. "Wasted: How America Is Losing Up to 40 Percent of Its Food from Farm to Fork to Landfill." National Resources Defense Council issue paper, 12-06-B. August 2012. http://www.nrdc.org/food/files/wasted-food-ip.pdf.

Gustavsson, Jenny, Christel Cederberg, Ulf Sonesson, Robert van Otterdijk, Alexandre Meybeck. "Global Food Losses and Food Waste: Extent, Causes and Prevention." Food and Agriculture Organization of the United Nations. Rome, 2011. http://www.fao.org/docrep/014/mb060e/mb060e.pdf.

Hawken, Paul, Amory B. Lovins, and L. Hunter Lovins. *Natural Capitalism: The Next Industrial Revolution.* 10th ed. Washington, DC: Earthscan, 2010.

Hayes, Denis. *Rays of Hope: The Transition to a Post-Petroleum World.* New York: W. W. Norton, 1977.

Hesterman, Oran B. *Fair Food: Growing a Healthy, Sustainable Food System for All.* New York: PublicAffairs, 2011.

Hillstrom, Kevin. *U.S. Environmental Policy and Politics: A Documentary History*. Washington, DC: CQ Press, 2010.

Holmgren, David. *Permaculture: Principles and Pathways beyond Sustainability*. Hepburn, Victoria: Holmgren Design Services, 2011.

Hyman, Leonard S., Andrew S. Hyman, and Robert C. Hyman. *America's Electric Utilities: Past, Present and Future*. 8th ed. Vienna, VA: Public Utilities Reports, 2005.

Intergovernmental Panel on Climate Change. *Climate Change 2007: Synthesis Report*. Contribution of Working Groups I, II and III to the Fourth Assessment Report of the Intergovernmental Panel on Climate Change. IPCC Plenary XXVII, Valencia, Spain, November 12–17, 2007. http://www.ipcc.ch/pdf/assessment-report/ar4/syr/ar4_syr.pdf.

Intergovernmental Panel on Climate Change. *Climate Change 2007: Synthesis Report—Summary for Policymakers*. Summary of the IPCC Plenary XXVII, Valencia, Spain, November 12–17, 2007. https://www.ipcc.ch/pdf/assessment-report/ar4/syr/ar4_syr_spm.pdf.

Jackson, Tim. *Prosperity without Growth: Economics for a Finite Planet*. London: Earthscan, 2011.

Jones, Van. *The Green-Collar Economy: How One Solution Can Fix Our Two Biggest Problems*. New York: HarperOne, 2008.

Kellert, Steven R. *Building for Life: Designing and Understanding the Human-Nature Connection*. Washington, DC: Island Press, 2005.

Kellert, Steven R., and Edward O. Wilson, eds. *The Biophilia Hypothesis*. Washington, DC: Island Press, 1993.

Kimbrell, Andrew, ed. *Fatal Harvest: The Tragedy of Industrial Agriculture*. Washington, DC: Foundation for Deep Ecology with Island Press, 2002.

Koplan, Jeffrey P., Catharyn T. Liverman, Vivica I. Kraak, eds. *Preventing Childhood Obesity: Health in the Balance*. Washington, DC: National Academies Press, 2005.

Leiserowitz, Anthony, Edward Maibach, Connie Roser-Renouf, Geoff Feinberg, and Peter Howe. "Global Warming's Six Americas, September 2012." Yale Project on Climate Change Communication and George Mason University Center for Climate Change Communication. New Haven, CT: Yale University and George Mason University, 2013. http://environment.yale.edu/climate-communication-OFF/files/Six-Americas-September-2012.pdf.

Lear, Linda J., ed. *Lost Woods: The Discovered Writing of Rachel Carson.* Boston: Beacon Press, 1998.

Lear, Linda J. *Rachel Carson: Witness for Nature.* 2nd ed. Boston: Mariner Books, 2009.

Lipscomb, Todd. *Re-Made in the USA: How We Can Restore Jobs, Retool Manufacturing, and Compete with the World.* Hoboken, NJ: Wiley, 2011.

Lopez, Anthony, Billy Roberts, Donna Heimiller, Nate Blair, and Gian Porro. *U.S. Renewable Energy Technical Potentials: A GIS-Based Analysis.* Technical Report NREL/TP-6A20-51946, July 2012. Golden, CO: National Renewable Energy Laboratory. http://www.nrel.gov/docs/fy12osti/51946.pdf.

Louv, Richard. *Last Child in the Woods: Saving Our Children from Nature-Deficit Disorder.* Chapel Hill, NC: Algonquin Books of Chapel Hill, 2008.

Lovejoy, Thomas E., and Lee Jay Hannah. *Climate Change and Biodiversity.* New Haven, CT: Yale University Press, 2005.

Lovins, Amory B. *Soft Energy Paths: Toward a Durable Peace.* San Francisco, CA: Friends of the Earth International, dist. Ballinger, 1977.

Lovins, Amory B., and the Rocky Mountain Institute. *Reinventing Fire: Bold Business Solutions for the New Energy Era.* White River Junction, VT: Chelsea Green, 2011.

Maathai, Wangari. *Unbowed: A Memoir.* New York: Knopf, 2006.

MacDicken, K., Ö. Jonsson, L. Piña, S. Maulo, Y. Adikari, M. Garzuglia, E. Lindquist, G. Reams, and R. D'Annunzio. *Global Forest Resources Assessment 2015: How Are the World's Forests Changing?* Rome: Food and Agriculture Organization of the United Nations, 2015. http://www.uncclearn.org/sites/default/files/inventory/a-i4793e.pdf.

Markowitz, Gerald E., and David Rosner. *Deceit and Denial: The Deadly Politics of Industrial Pollution.* Berkeley: University of California Press, 2013.

Martinez, Steve, Michael Hand, Michelle Da Pra, Susan Pollack, Katherine Ralston, Travis Smith, Stephen Vogel, Shellye Clark, Luanne Lohr, Sarah Low, and Constance Newman. *Local Food Systems: Concepts, Impacts, and Issues.* Economic Research Report 97. Washington, DC: US Department of Agriculture, Economic Research Service, May 2010. http://www.ers.usda.gov/media/122868/err97_1_.pdf.

McCollester, Charles. *The Point of Pittsburgh: Production and Struggle at the Forks of the Ohio.* Pittsburgh, PA: Battle of Homestead Foundation, 2008.

McCraw, Thomas K. *Prophets of Regulation: Charles Francis Adams, Louis D.*

Brandeis, James M. Landis, Alfred E. Kahn. Cambridge, MA: Belknap Press, 1984.

McDonough, William, and Michael Braungart. *Cradle to Cradle: Remaking the Way We Make Things*. New York: North Point Press, 2002.

Meadows, Dennis L., William W. Behrens II, Donella H. Meadows, Roger F. Naill, Jørgen Randers, Erich K. O. Zahn, et al. *Dynamics of Growth in a Finite World*. Cambridge, MA: Wright-Allen Press, 1974.

Meadows, Donella, Jørgen Randers, Dennis Meadows. *Limits to Growth: The 30-Year Update*. White River Junction, VT: Chelsea Green, 2004.

Millennium Ecosystem Assessment Board. *Ecosystems and Human Well-Being: Synthesis*. Washington, DC: Island Press, 2005. http://www.millenniumassessment.org/documents/document.356.aspx.pdf.

Moore, Robin C., and Herbert H. Wong. *Natural Learning: The Life History of an Environmental Schoolyard*. Berkeley, CA: MIG Communications, 1997.

National Science Board. "Science and Technology: Public Attitudes and Understanding." *Science and Engineering Indicators 2014*. Washington, DC: National Science Foundation, 2014. http://www.nsf.gov/statistics/seind14/.

Nellemann, Christian, and Emily Corcoran, eds. *Dead Planet, Living Planet: Biodiversity and Ecosystem Restoration for Sustainable Development: A Rapid Response Assessment*. Nairobi, Kenya: United Nations Environment Programme, 2010.

Nellemann, Christian, Monika MacDevette, Ton Manders, Bas Eickhout, Birger Svihus, Anne Gerdien Prins, Bjørn P. Kaltenborn, eds. *The Environmental Food Crisis: The Environment's Role in Averting Future Food Crises: A UNEP Rapid Response Assessment*. Arendal, Norway: United Nations Environment Programme, 2009.

Nixon, Rob. *Slow Violence and the Environmentalism of the Poor*. Cambridge, MA: Harvard University Press, 2011.

Odum, Eugene P. *Fundamentals of Ecology*. 2nd ed. Philadelphia, PA: W. B. Saunders, 1966.

Oreskes, Naomi, and Erik M. Conway. *Merchants of Doubt: How a Handful of Scientists Obscured the Truth on Issues from Tobacco Smoke to Global Warming*. New York: Bloomsbury, 2010.

Office of the United Nations High Commissioner for Human Rights. *Convention on the Rights of the Child*. United Nations General Assembly Resolution 44/25. November 20, 1989. http://www.un.org/documents/ga/res/44/a44r025.htm.

Pew Research Center. "How Americans View Top Energy and Environmental Issues." *Pew Research Center*. January 15, 2015. Accessed June 1, 2015. http://www.pewresearch.org/key-data-points/environment-energy-2/.

Phillips, Charles F., Jr. *The Regulation of Public Utilities: Theory and Practice*. Arlington, VA: Public Utilities Reports, 1985.

Pimentel, David, P. Hepperly, James Hanson, David Douds, and Rita Seidel. "Environmental, Energetic, and Economic Comparisons of Organic and Conventional Farming Systems." *BioScience* 55, no. 7 (2005): 573–82.

Rifkin, Jeremy. *The Third Industrial Revolution: How Lateral Power Is Transforming Energy, the Economy, and the World*. New York: Palgrave Macmillan, 2011.

Rodale, Robert. *Regenerative Organic Agriculture and Climate Change: A Down-to-Earth Solution to Global Warming*. Kutztown, PA: Rodale Institute, 2014. http://rodaleinstitute.org/assets/RegenOrgAgricultureAndClimateChange_20140418.pdf.

Robèrt, Karl-Henrick. "Tools and Concepts for Sustainable Development, How Do They Relate to a Framework for Sustainable Development, and to Each Other?" *Journal of Cleaner Production* 8, no. 3 (2000): 243–54.

Rosegrant, Mark W., and Siwa Msangi. "Consensus and Contention in the Food-Versus-Fuel Debate." *Annual Review of Environment and Resources* 39, no. 1 (2014): 271–94.

Sachs, Jeffrey D. *Common Wealth: Economics for a Crowded Planet*. New York: Penguin, 2008.

Sandler, Ronald, and Philip Cafaro, eds. *Environmental Virtue Ethics*. Lanham, MD: Rowman and Littlefield, 2005.

Savitz, Andrew W., and Karl Weber. *The Triple Bottom Line*. San Francisco, CA: Jossey-Bass, 2006.

Schmidt, Michael R. *Performance-Based Ratemaking: Theory and Practice*. Vienna, VA: Public Utilities Reports, 2000.

Searchinger, Tim, Craig Hanson, Janet Ranganathan, Brian Lipinski, Richard Waite, Robert Winterbottom, Ayesha Dinshaw, and Ralph Heimlich. *Creating a Sustainable Food Future: A Menu Of Solutions to Sustainably Feed More than 9 Billion People by 2050*. World Resources Report 2013–14: Interim Findings. Washington, DC: World Resources Institute, 2014. http://www.unep.org/gpa/documents/publications/CreatingSustainableFoodFuture.pdf.

Secretariat of the Convention on Biological Diversity. *Global Biodiversity Out-*

look 3. Montreal: Secretariat of the Convention on Biological Diversity, 2010. https://www.cbd.int/doc/publications/gbo/gbo3-final-en.pdf.

Schug,T. T., R. Abagyan, B. Blumberg, T. J. Collins, D. Crews, P. L. DeFur, S. M. Dickerson, T. M. Edwards, A. C. Gore, L. J. Guillette, T. Hayes, J. J. Heindel, A. Moores, H. B. Patisaul, T. L. Tal, K. A. Thayer, L. N. Vandenberg, J. C. Warner, C. S. Watson, F. S. vom Saal, R. T. Zoeller, K. P. O'Brien, J. P. Myers. "Designing Endocrine Disruption Out of the Next Generation of Chemicals." *Green Chemistry* 15, no. 1 (2013): 181–98.

Silverstein, Ken. "Creative Disruption." *Public Utilities Fortnightly*. April 2015, 7.

Smith, Lindsay P., Shu Wen Ng, and Barry M. Popkin. "Trends in US Home Food Preparation and Consumption: Analysis of National Nutrition Surveys and Time Use Studies from 1965–1966 to 2007–2008." *Nutrition Journal* 12 (2013): 45.

Sobel, David. *Beyond Ecophobia: Reclaiming the Heart of Nature Education*. Great Barrington, MA: Orion Society, 1996.

Solomon, Susan, Dahe Qin, Martin Manning, Melinda Marquis, Kristen Averyt, Melinda M. B. Tignor, and Henry LeRoy Miller, Zhenlin Chen, eds. *Climate Change 2007: The Physical Science Basis*. Contribution of Working Group I to the Fourth Assessment Report of the Intergovernmental Panel on Climate Change. Cambridge: Cambridge University Press, 2007. http://www.ipcc.ch/pdf/assessment-report/ar4/wg1/ar4_wg1_full_report.pdf.

Spencer, Christopher, and Mark Blades, eds. *Children and Their Environments: Learning, Using, and Designing Spaces*. Cambridge, UK: Cambridge University Press, 2006.

Srinivasan, Shobha, Liam R. O'Fallion, Allen Dearry. "Creating Healthy Communities, Healthy Homes, Healthy People: Initiating a Research Agenda on the Built Environment and Public Health." *American Journal of Public Health* 93, no. 9 (2003): 1446–50.

Stanley, Steven M. *Earth System History*. New York: W. H. Freeman, 1999.

Stark, Camila, Jacquelyn Pless, Jeffrey Logan, Ella Zhou, and Douglas J. Arent. *Renewable Electricity: Insights for the Coming Decade*. Technical report NREL/TP-6A50-63604. Golden, CO: Joint Institute for Strategic Energy Analysis, February 2015.

Suzuki, David. *Sacred Balance: Rediscovering Our Balance in Nature*. Rev. ed. Vancouver: Greystone Books, 2007.

Tarr, Joel A., ed. *Devastation and Renewal: An Environmental History of Pittsburgh and Its Region*. Pittsburgh, PA: University of Pittsburgh Press, 2003.

Thevenon, Florian, Chris Carroll, and João Sousa, eds. *Plastic Debris in the Ocean: The Characterization of Marine Plastics and Their Environmental Impacts, Situation Analysis Report*. Gland, Switzerland: International Union for Conservation of Nature and Natural Resources, 2014. https://portals.iucn.org/library/sites/library/files/documents/2014-067.pdf.

Thomas, Mary Adam. *Building in Bloom*. Portland, OR: Ecotone Publishing, 2013.

Thompson, Allen, and Jeremy Bendick-Keymer. *Ethical Adaptation to Climate Change: Human Virtues of the Future*. Cambridge, MA: MIT Press, 2012.

United Nations Department of Economic and Social Affairs. *World Population Prospects: The 2002 Revision*. Volume 3, *Analytical Report*. New York: United Nations, 2004. http://www.un.org/esa/population/publications/wpp2002/WPP2002_VOL_3.pdf.

United Nations Department of Economic and Social Affairs, Population Division. *World Population Prospects: The 2015 Revision. Key Findings and Advance Tables*. Working paper no. ESA/P/WP.241. New York: United Nations, 2015. https://esa.un.org/unpd/wpp/publications/files/key_findings_wpp_2015.pdf.

United Nations Framework Convention on Climate Change. *Report of the Conference of the Parties on Its Twenty-First Session, Held in Paris from 30 November to 15 December, 2015*. Paris: United Nations, 2016. http://unfccc.int/resource/docs/2015/cop21/eng/10.pdf.

United Nations Framework Convention on Climate Change. Adoption of the Paris Agreement, Draft Decision/CP21. Paris: United Nations, 2015. https://unfccc.int/resource/docs/2015/cop21/eng/l09r01.pdf.

United States Department of Agriculture, National Agricultural Statistics Service. *Farms, Land in Farms, and Livestock Operations 2012 Summary*. Washington, DC: US Department of Agriculture, February 2013. http://usda.mannlib.cornell.edu/usda/nass/FarmLandIn/2010s/2013/FarmLandIn-02-19-2013.pdf.

United States Department of Agriculture, Economic Research Service Statistics. *Adoption of Genetically Engineered Crops in the U.S.* Washington, DC: United States Department of Agriculture, 2015. http://www.ers.usda.gov/data-products/adoption-of-genetically-engineered-crops-in-the-us.aspx.

United States Department of Health and Human Services, Centers for Disease Control and Prevention. *Fourth National Report on Human Exposure to Environmental Chemicals, 2009*. Atlanta, GA: Department of Health and Human

Services, Centers for Disease Control and Prevention, December 2009. http://www.cdc.gov/exposurereport/pdf/fourthreport.pdf.

United States Environmental Protection Agency. "Future of Climate Change." *Climate Change Science.* Accessed May 10, 2016. https://www3.epa.gov/climatechange/science/future.html.

United States Environmental Protection Agency. *Inventory of U.S. Greenhouse Gas Emissions and Sinks: 1990–2013.* Washington, DC: US Environmental Protection Agency, 2015. https://www3.epa.gov/climatechange/Downloads/ghgemissions/US-GHG-Inventory-2015-Main-Text.pdf.

United States Environmental Protection Agency. "TRI-Listed Chemicals." *Toxic Release Inventory (TRI) Program.* Accessed June 1, 2015. http://www2.epa.gov/toxics-release-inventory-tri-program/tri-listed-chemicals 2015.

Vaughan-Lee, Llwellyn, ed. *Spiritual Ecology: The Cry of the Earth.* Point Reyes, CA: Golden Sufi Center, 2013.

Vince, Gaia. *Adventures in the Anthropocene: A Journey to the Heart of the Planet We Made.* Minneapolis, MN: Milkweed Editions, 2014.

Waage, Sissel, ed. *Ants, Galileo, and Gandhi: Designing the Future of Business through Nature, Genius, and Compassion.* Sheffield, UK: Greenleaf Publishing, 2003.

Ward, Barbara, and René J. Dubos. *Only One Earth: The Care and Maintenance of a Small Planet.* New York: W. W. Norton, 1972.

Weidensaul, Scott. *Living on the Wind: Across the Hemisphere with Migratory Birds.* New York: North Point Press, 1999.

Williams, Eric, and Ruediger Kuehr, eds. *Computers and the Environment: Understanding and Managing their Impacts.* Dordrecht: Kluwer Academic Publishers, 2003.

Wilson, Edward O. *The Diversity of Life.* Cambridge, MA: Harvard University Press, 1992.

Wilson, Edward O. *Naturalist.* Washington, DC: Island Press, 1994.

Wilson, Edward O. *The Social Conquest of Earth.* New York: Liveright, 2012.

Winston, Mark L. *Nature Wars: People vs. Pests.* Cambridge, MA: Harvard University Press, 1997.

Wood, Mary Christina. *Nature's Trust: Environmental Law for a New Ecological Age.* New York: Cambridge University Press, 2014.

World Commission on the Ethics of Scientific Knowledge and Technology. *The Ethical Implications of Global Climate Change: Report by the World Commis-*

sion on the Ethics of Scientific Knowledge and Technology. Paris: United Nations Educational, Scientific and Cultural Organization, 2010.

World Commission on the Ethics of Scientific Knowledge and Technology. *The Precautionary Principle*. Paris: United Nations Educational, Scientific and Cultural Organization, 2005. http://unesdoc.unesco.org/images/0013/001395/139578e.pdf.

Worldwatch Institute. *State of the World 2014: Governing for Sustainability*. Washington, DC: Island Press, 2014.

INDEX

Note: Page numbers in italics refer to figures or tables.